ALSO BY STEPHEN S. HALL

Invisible Frontiers:
The Race to Synthesize a Human Gene

Mapping the Next Millennium:
The Discovery of New Geographies

A Commotion in the Blood: Life, Death,
and the Immune System

Merchants of Immortality:
Chasing the Dream of Human Life Extension

Size Matters: How Height Affects the Health, Happiness,
and Success of Boys—and the Men They Become

WISDOM

WISDOM

FROM PHILOSOPHY
TO NEUROSCIENCE

STEPHEN S. HALL

Alfred A. Knopf　NEW YORK · 2010

THIS IS A BORZOI BOOK
PUBLISHED BY ALFRED A. KNOPF

Copyright © 2010 by Stephen S. Hall

All rights reserved. Published in the United States by Alfred A. Knopf,
a division of Random House, Inc., New York, and in Canada
by Random House of Canada Limited, Toronto.
www.aaknopf.com

Owing to limitations of space, all acknowledgments to reprint previously
published material may be found at the end of the volume.

Knopf, Borzoi Books, and the colophon are
registered trademarks of Random House, Inc.

Library of Congress Cataloging-in-Publication Data
Hall, Stephen S.
Wisdom : from philosophy to neuroscience / by Stephen S. Hall—1st ed.
p.   cm.
"A Borzoi book."
Includes bibliographical references and index.
ISBN 978-0-307-26910-2 (hardcover : alk. paper)   1. Wisdom.   2. Neuropsychology
I. Title.
BJ1595.H28 2010
179'.99—dc22   2009027438

Manufactured in the United States of America

First Edition

for Robert S. Hall

(1926–2008)

"A word to the wise . . ."

and

Anne Friedberg

(1952–2009)

Man has two windows to his mind: through one he can see his own self as it is; through the other, he can see what it ought to be. It is our task to analyse and explore the body, the brain and the mind of man separately; but if we stop here, we derive no benefit despite our scientific knowledge. It is necessary to know about the evil effects of injustice, wickedness, vanity and the like, and the disaster they spell where the three are found together. And mere knowledge is not enough; it should be followed by appropriate action. An ethical idea is like an architect's plan.

—Gandhi

CONTENTS

PART THREE: BECOMING WISE

WISDOM DEFINED
(SORT OF)

You, my friend . . . are you not ashamed . . . to care so little about wisdom and truth and the greatest improvement of the soul, which you never regard or heed at all?

—Socrates, defending himself at his trial

CHAPTER ONE

WHAT IS WISDOM?

The days of our life are seventy years,
or perhaps eighty if we are strong;
even then their span is only toil and trouble;
they are soon gone, and we fly away . . .
So teach us to count our days
that we may gain a wise heart.

—Psalm 90

That man is best who sees the truth himself,
Good too is he who listens to wise counsel.
But who is neither wise himself nor willing
To ponder wisdom is not worth a straw.

—Hesiod

ON A BEAUTIFUL FALL MORNING nearly a decade ago, like hundreds of mornings before and since, I dropped off one of my children at school. Micaela, then five years old, had just started first grade, and the playground chatter among both the children and their parents reflected that mix of nervous unfamiliarity and comforting reconnection that marks the beginning of the school year. I lingered in the schoolyard until Micaela lined up with her teacher and classmates. She wore a pretty purple dress that my mother had just sent her, white socks, and pink-and-white-checkered sneakers. A hair band exposed her hopeful, eager, beautiful face. I sneaked in a last hug, as impulsive dads are wont to do, before she disappeared into the building. The time was about 8:40 a.m.

As I left the schoolyard and began to head toward the subway and home to Brooklyn, I heard a thunderous, unfamiliar roar overhead. As

3

the noise grew louder and closer, I froze in an instinctive crouch, much like the rats we always read about in scientific experiments on fear, wondering where the sound was coming from, knowing only that it was ominously out of the ordinary. Moments later, a huge shadow with metal wings passed directly over my head, like some prehistoric bird of prey. I instantly recognized it as a large twin-engine commercial airliner, but nothing in my experience prepared me for what happened next. I watched for the endless one . . . two . . . three . . . four seconds it took for this shiny man-made bird to fly directly into the tall building that I faced several blocks away. In real time, I watched a 395,000-pound airplane simply disappear. Almost immediately black smoke began to curl out of the cruel, grinning incision its wings had sliced in the façade of the skyscraper.

In moments when life's regular playbook flies out the window, when the ground shifts beneath our feet in a literal or figurative earthquake, we feel a surge of adrenalized fear at the shock of the unexpected. But right behind that feeling comes the struggle to make sense of the seemingly senseless, to try to understand what has just happened and what it means so that we will know how to think about a future that suddenly seems uncertain and unpredictable. In truth, the future is *always* unpredictable, which is why these moments of shock remind us, with unusual urgency, that we have a constant (if often unconscious) need for wisdom, too.

Although we now all know exactly what happened that terrible morning, the ground truth in lower Manhattan on September 11, 2001, was much fuzzier at 8:45 a.m. One of the hallmarks of wisdom, what distinguishes it so sharply from "mere" intelligence, is the ability to exercise good judgment in the face of imperfect knowledge. In short, do the right thing—ethically, socially, familiarly, personally. Sometimes, as on this day, we have to deliberate these decisions in the midst of an absolutely roaring neural stew of conscious and unconscious urgings. In one sense, I knew exactly what had happened long before the first news bulletin hit the airwaves. In a larger sense, someone watching television in Timbuktu soon knew vastly more about the big picture than I did. This may be an exaggerated example, but it is in precisely the murk of this kind of confusion that we often have to make decisions. So what did I do?

I went to a nearby shop and bought a cup of coffee.

It didn't occur to me until much later that this was a decision of sorts—perhaps a foolish one, and certainly not an obvious one. But to the extent that I mustered even a dram of wisdom that day, it was in how I viewed the situation and what I thought was most important. Oddly, I felt little or no physical threat, despite such close proximity to the unfolding disaster; in some respects, the event played much scarier on TV than in person. My immediate focus, even then, was on the long-term psychological impact that such a calamity might have on a young child, and what (if anything) a parent might do to minimize it. I hadn't quite understood yet that that would be my mission for the day, but by standing in the street and sipping a cup of coffee, in that mysterious shorthand of human choice, I had chosen to stay close to my daughter, to stay calm, and, failing that, to fake parental calm realistically enough to convince her that this was a situation we could deal with.

But she didn't need to see the whole movie. I did not think it was a good idea for a young child to witness, as I did, human bodies falling like paperweight angels from the upper floors of the nearby tower. Even more, I did not think it was a good idea for a young child to absorb, even for a moment, the panic and despair written on the faces of all the adults who were beginning to comprehend that the world as they had known it, even a few minutes earlier, had suddenly changed, slipping irrevocably out of their (however illusionary) controlling grasp.

If you're thinking that I'm offering a smug little narrative about wise parenting, not to worry. Wisdom doesn't come easily to us mortals, and I've been reminded many times since that it probably didn't come to me that day, either. Many of the choices I made that morning were second-guessed by my wife, by my friends, and even by my daughter. More to the point, my small-minded plan to buffer Micaela's emotional experience was rudely interrupted by the collapse of 500,000 tons of metal, concrete, and glass. Just as teachers began to evacuate children from the school, the second tower came down, unleashing the kind of apocalyptic roar no child should ever have to hear, and a huge pyroclastic cloud of debris came boiling up Greenwich Street toward us. You couldn't tell if the cloud was going to reach us or not, but it wasn't a moment for contemplation. I picked up Micaela and we joined a horde of people running up the street. As I carried her in my arms, swimming upriver in a school of panicked fish, she was forced to look backward, downtown, right into the onrushing menace of our suddenly dark times. Even to this day, however, the thing Micaela remembers most about the evacuation is the moment her classmate Liam accidentally walked into a street sign when he wasn't looking.

It will be a long time, if ever, before I know if I acted wisely on 9/11. Indeed, it didn't even occur to me until I was writing this passage that the most important decision I made that day did not even rise to the level of conscious choice. I "decided," without any conspicuous deliberation, that I had to be a parent first, not a journalist, on that particular morning. At one level, it was an obvious choice; at another, it went against self-interest, career, my professional identity, taking advantage of being an eyewitness to the biggest story of my lifetime. What was I thinking?

That, in a sense, is what I want this book to be about: How do we make complex, complicated decisions and life choices, and what makes some of these choices so clearly wise that we all intuitively recognize them as a moment, however brief, of human wisdom? What goes on in our heads when we're struggling to be patient and prudent, and are there ways to enhance those qualities? When we're being foolish, on the other hand, do our brains make us do it? And how does the passage of time, and our approaching mortality, change our thought processes and perhaps make us more amenable to wisdom?

In moments of exceptional challenge and uncertainty, we tend to ask, How did this happen? What could we have done to prevent this dire turn of events? This is another way of saying, I realize now, that we are always searching for wisdom, but all too often we are looking for it in the rearview mirror, sifting the past for clues to how we might have thought about the future in a different way.

We crave wisdom—worship it in others, wish it upon our children, and seek it ourselves—precisely because it will help us lead a meaningful life as we count *our* days, because we hope it will guide our actions as we step cautiously into that always uncertain future. At times of challenge and uncertainty, nothing seems more important than wisdom—economic wisdom, moral wisdom, political wisdom, even that private, behind-closed-doors wisdom that allows us to convey the gravity of changed circumstances to our children without making them afraid of change itself.

Nothing seems more important, yet nothing seems more beyond our grasp, until we begin to think about wisdom before we think we need it.

I am not an expert on wisdom (in the most important sense, none of us is). I'm just a journalist who for many years has written about science, which in some circles even further disqualifies me from having anything of value to say about wisdom. But all of us find ourselves in situations that demand it, and we don't need a 9/11 or a cataclysmic economic collapse to bring our desire for wisdom front and center. A car accident, the loss of a job, sudden illness, a floundering relationship, deep disagreements with parents or children—any old run-of-the-mill crisis will do.

We all aspire to have wisdom. Not necessarily because it will guarantee us happier, more fulfilling, better lives (although those have been worthy goals almost from the moment philosophers began to contemplate it), but because wisdom *as a process* can serve as a guide to helping us make the best-possible decisions at junctures of great importance in our lives. With an added, implicit (or sometimes explicit) tincture of mortality, it can get us to slow down long enough to think about actions and consequences. It can help us frame problems in a different way, allowing us to see unexpected solutions. It can help us maximize the good we do not only in the intimate community of family and friends but also in the larger communities that define our social identity as neighbors, residents, citizens, congregants, and custodians of the planet.

Many of these decisions are years in the planning and preparation, like selecting a mate or choosing a career. Some of them arrive with the roar of a hijacked plane or the suddenness of a phone call from the doctor. At the same time, we can't separate those crossroads moments from the "vehicle," the lifetime of experiences, that brought us to the intersection in the first place. Was this vehicle well maintained? Was it tested in all sorts of emotional weather, on every kind of situational terrain? Wisdom resides not just in the decision per se but also, as Confucius perhaps best of all philosophers shrewdly understood, in the Way of life—what he called *gen*—that precedes the decision.

Decision making lies at the heart of wisdom, but it's not the whole story. Making those decisions, in turn, draws on a subtle weave of intellectual, emotional, and social gifts—gathering information, discerning the reality behind artifice (especially when it comes to human nature), evaluating and editing that accumulated knowledge, listening to one's heart and one's head about what is morally right and socially just, thinking not only of oneself but others, thinking not only in the here and now but about the future. Even in times of crisis, however, wisdom sometimes demands the paradoxical decision to resist doing something just for the sake of doing it—that flailing impulse "to do something, *anything*" that social scientists sometimes call the "action bias." "Some of the wisest and most devout men," the French essayist and philosopher Michel de Montaigne observed, "have lived avoiding all noticeable actions."

If wisdom weren't important, no one would even bother arguing about its definition. But that's the point: It *is* important, and every one of us, because we do lead lives and want those lives to be as good as they can be, is, to a certain extent, an expert in wisdom, even if (as is certainly the case with almost all of us) it is an expertise grounded in want, not possession. All of us have an intuitive sense of what wisdom means and what constitutes wise behavior. In a rough, nonacademic sense (to paraphrase Supreme Court Justice Potter Stewart's famous opinion about pornography), we know it when we see it, even if we can't define it.

That may suffice as a satisfyingly casual approach to personal philosophy, but such definitional squishiness usually makes for bad science, and this is, in many ways, a book about science's improbable exploration (if not annexation) of one of philosophy's most prized duchies. No one in a modern laboratory would argue that wisdom is a tractable subject for research; many scientists reasonably view it as something like intellectual libel to suggest that experiments in their labs have anything whatsoever to do with such a fuzzy topic. Even social scientists have trodden lightly; Paul B. Baltes, who probably studied wisdom with more depth and empirical rigor than any other psychologist in the modern era, spoke of a "fuzzy zone" of wisdom, where human expertise never quite rises to an idealized level of knowledge about the human condition.

But the struggle to define wisdom is embedded in the texture of its philosophical, psychological, and cultural history. And every time we think about it, every time we make the mighty effort to pause and contemplate a potential role for wisdom in whatever we are about to do or say, we join that noble struggle and move a step closer to achieving it. In trying to define wisdom, we are not merely engaging in a dry academic exercise. We are, in a fundamental and indeed essential sense, engaging in a conversation with ourselves about how to lead the best-possible life. We are engaging in a conversation with ourselves about who we want to be by the time we complete that journey and, in the words of Psalm 90, "fly away."

Wisdom begins with awareness, of the self and the world outside the self; it deepens with our awareness of the inherent tension between the inner "I" and the outer world.

I began to realize this when I was asked to write an article for *The New York Times Magazine* about wisdom research—or, as the cover line asked, "Can Science Tell Us Who Grows Wiser?" As I quickly discovered, there's no shortage of definitions of wisdom, and no dearth of disagreement about them; in an academic anthology entitled *Wisdom: Its Nature, Origins, and Development,* published in 1990, there are thirteen separate chapters written by prominent psychologists, and each one offers a different definition of wisdom. As Robert J. Sternberg succinctly put it, "To understand wisdom fully and correctly probably requires more wisdom than any of us have."

But *thinking* about wisdom nudges us closer to the thing itself. Every time I encountered a new definition of wisdom, or some argument from the psychological literature, I found myself considering my own life: my decisions, my values, my shortcomings, my choices in confronting difficult practical and moral dilemmas. If some psychologist had identified emotional evenhandedness as a component of wisdom, I would pause to consider my own emotional behavior. What set me off emotionally, and what kinds of decisions did I make—things said, actions taken, tone of voice and physical vocabulary—when I had to deal, for example, with a frustrating situation with professional colleagues or with my children's inconvenient moments of emotional demand? When compassion emerged as a central component, I was forced to consider the limitations and inconsistencies of my own behavior. When I read the work of Baltes, who believed that dealing with ambiguity and uncertainty was a central aspect of modern wisdom, I realized that moments of ambiguity and uncertainty are often the most stressful and challenging of our lives. (This form of self-consciousness reminded me of the way I used to obsess about diseases when I wrote about medical conditions, but this was more like a philosophical form of hypochondria, much less scary and much more illuminating.) With each new question, I realized that I had unwittingly embarked on an impromptu program of mental exercise, an informal calisthenics of self-awareness.

As I burrowed deeper into the literature of wisdom, I found myself silently mouthing the same question over and over to myself whenever I confronted a problem or dilemma: What would be the wisest thing to do here? I won't say I *acted* wisely—as Baltes and many others have pointed out, wisdom is more an ideal aspiration than a state of mind or a pattern of behavior that we customarily inhabit. But simply framing a decision in those terms was intellectually and emotionally bracing. I came away from this experience discovering (in the process of researching and writing a brief magazine article) that as soon as you are confronted with a definition of wisdom, however provisional or tentative, however debatable or howlingly inadequate, you are forced to view that definition through the prism of your own history and experience. Which is another way of saying that we all have a working definition of wisdom floating around in our heads, but we are rarely forced to consider it, or consult it, or challenge it, or amend it, much less apply any standard of wisdom to gauge our own behavior and decisions on a daily basis.

Simply put, thinking about wisdom forces you to think about the way you lead your life, just as reading about wisdom, I believe, forces you to wrestle with its meaning and implications. You might come to think of this exercise, as I have, as an enlightened form of self-consciousness, almost an armchair form of mindfulness or meditation that cannot help but inform our actions. And that's another key point: to separate wisdom from action is a form of malpractice in the conduct of one's life. "We ought to seek out virtue not merely to contemplate it," Plutarch wrote, "but to derive benefit from doing so."

Soon, whenever I found myself in a challenging situation—refereeing a sibling spat, confronting interpersonal friction with a loved one or friend, being called upon to deal with something that triggered titanic forces of procrastination, or even weighing a trivial dilemma of daily compassion, such as deciding whether to give a poor person some spare change—I felt myself slowing down long enough to ask myself that question: What would be the wisest thing to do? I realize this was very

small potatoes compared to Mother Teresa working in the slums o
cutta or Martin Luther King, Jr., marching on Selma, and I won't
did this all the time—a conscientiously wise person might easily ex
ence an existential form of rigor mortis, paralyzed by serial episode. or
deliberation.

But I found it a refreshing exercise. It forced me to clarify choices. It
slowed down the clock of urgency against which we all seem to be racing
as we struggle with decisions. It allowed me to step outside of myself and
momentarily stifle the urges of my innate selfishness—second to none, I
submit, yet probably pretty much equivalent to everybody else's—long
enough to see a bigger picture. It had an archaic but familiar quality of
self-monitoring. It felt, for lack of a better word, *responsible*—not in the
sense that others hold us responsible, but, rather, in terms of raising the
bar of expectations we hold for ourselves.

But what exactly do I mean by wisdom?

Many definitions of wisdom converge on recurrent and common elements: humility, patience, and a clear-eyed, dispassionate view of human
nature and the human predicament, as well as emotional resilience, an
ability to cope with adversity, and an almost philosophical acknowledgment of ambiguity and the limitations of knowledge. Like many big
ideas, it's also nettled with contradictions. Wisdom is based upon knowledge, but part of the physics of wisdom is shaped by uncertainty. Action
is important, but so is judicious inaction. Emotion is central to wisdom,
yet emotional detachment is indispensable. A wise act in one context
may be sheer folly in another.

These inherent contradictions do not fatally vex a potential definition
of wisdom; rather, they are embedded in it. One of the best ways to
think about wisdom, in fact, is to try to identify those rare individuals
who manage to reconcile these contradictions and still embody wisdom.
These are (or once were) living, breathing, and, because they are human,
imperfect definitions of wisdom, but they are also less abstract, more like
wisdom in the flesh. We can learn a lot about wisdom from its exemplars, past and present.

A few years ago, psychologists in Canada conducted a study in which
they asked subjects to nominate people, historical or modern, who
struck them as especially wise. There are plenty of problems with these

so-called questionnaire studies, beginning with the fact that they are typically inflicted on college undergraduates and may represent a narrow, undercooked demographic slice of wisdom pie. Nonetheless, consider the names they came up with: Mahatma Gandhi, Confucius, Jesus Christ, Martin Luther King, Jr., Socrates, Mother Teresa, Solomon, Buddha, the Pope, Oprah Winfrey, Winston Churchill, the Dalai Lama, Ann Landers, Nelson Mandela, and Queen Elizabeth II, in that order. Overwhelmingly historical (what does it say about contemporary culture that one of the few living exemplars of modern wisdom is a talk-show host?), predominantly male, surprisingly rich in social activists, and yet a reasonably wise catch for an admittedly porous cultural net. As with wisdom, we all seem to recognize wise people when we see them.

The fact that so many of the figures were historical adverts to humankind's enduring cultural fascination with the topic—wisdom is never out of fashion, and wise people speak to us beyond their time, place, and circumstance. The nomination of contemporary figures like Mr. Mandela and Ms. Winfrey, on the other hand, is a reassuring ratification of the notion that wisdom still exerts a strong cultural hold on the modern world, and reminds us that a core element of wisdom is the commitment to social justice and the greater public good. Many of us could quibble with some of the names on this Sage Hit Parade (indeed, you might provoke a very interesting conversation if you discussed the relative merits of these "wise" people at a dinner party). Most of us would nonetheless agree that they represent a reasonably exalted conclave of thoughtful individuals who could act wisely, at least part of the time.

Several startling facts leap out from this list, however. One is its dark message about the inherent threat posed by insistently wise behavior: In a profound sense, the figures we now celebrate for their wisdom often had a deeply adversarial relationship with the prevailing values of the societies in which they lived. Indeed, Pythagoras—who gave us the word *philosophy* (literally, "love of wisdom"), who identified three distinct "lifestyles" (the acquisitive, the competitive, and the contemplative), and who argued that contemplation, awareness, or, in the Eastern idiom, "awakening" is by far the best lifestyle—so alienated the populace of Croton, where he lived, that its citizens burned his house down, massacred many of his fellow Pythagoreans, and forced the "inventor" of philosophy to flee for his life. In an early foreshadowing of the Socratic tragedy, we can see that in many cultures, the wise man is also a marked man. Many of the wise people on the list needed to abandon conventional modes of life and thought to nurture the habit of mind for which they are now celebrated, which is often to tell society what it doesn't want to hear; many were ostracized during their lifetime, while others were executed outright or assassinated. Mandela and Gandhi were imprisoned; Confucius was unemployable; Socrates was put to death; even the closest friends of Jesus Christ, according to philosopher Karl Jaspers, viewed him as a madman. In its particular time and place, wisdom not only perturbs but often appears socially dangerous.

The other surprise, and a troubling one at that, is the relative absence of women. The presence of so many wise men is more than a gender aberration; it tells us something very important about our working definition. Wisdom clearly isn't a trait conferred by a gene sequestered on the male Y chromosome. For every Solomon, there is a Sarah and an Esther; for every Pericles, there is an Aspasia, his little-known mistress, who, according to Plutarch, was one of the wisest people in that wisest era of Greek civilization. For every Jesus, a Mary Magdalene; for every Mandela, an Aung San Suu Kyi. In the Hebrew Bible, wisdom is a She.

So why so few women? I don't think there's a dearth of female wisdom, just a painfully slow evolution in the cultural notion of wisdom and an equally painful and long-standing disenfranchisement of women from the public domain of wisdom for many centuries. The ancient Greeks personified wisdom in the goddess Athena, but at the same time, Athenian women were not citizens, could not speak or vote in the assembly, could not sit on juries, could not select their own marriage partners or the age at which they could marry. Did this deprive them of wisdom? Of course not. As a recent art exhibit ("Worshiping Women: Ritual and Reality in Classical Athens") made clear, the female deities of Athens—Athena, Artemis, Demeter, and Aphrodite—were all role models for a private, domestic, almost mystical domain of wisdom. Art critic Holland Cotter got it exactly right in noting, "Birth and death—the only real democratic experiences, existentially speaking—were in women's hands." In her strategic abstinence, Lysistrata is as much an exemplar of Athenian wisdom as Socrates; in her enfranchisement of emotion as parcel to thought, Sappho was closer to modern neuroscience than Plato.

So the relative dearth of wise women is not a problem in the stars, but in us: We need to be a little more catholic about where we look for, and are willing to find, wisdom. I don't mean we should throw out all the usual Old White Guys, in all the usual Great Books; they make very good company. We can find useful provocations about the meaning of wisdom in the dialogues of Plato and in the proverbs of the Bible, in the lamentations of Saint Augustine and especially in the clear-eyed though often grumpy insights of Montaigne, who once declared, "The most manifest sign of wisdom is continual cheerfulness." If by that he meant optimism about the future, he is backed up by neuroscientists, who have begun to find support for that notion.

But as one of the most famous Old White Guys observed, in the voice of Poor Richard and his humble bromides, "There have been as great Souls unknown to fame as any of the most famous." The truth is that we can find wisdom not only on the steps of the Parthenon, but also around a family dinner table; not only in the pages of the Everyman's Library but also in the funny pages (in a recent biography of cartoonist Charles Schulz, writer David Michaelis captured the minimalist wisdom of "Peanuts" when he described it as a comic strip "about people working out the interior problems of their daily lives without ever actually solving them"). The fact is that women have historically exerted their abundant wisdom out of the public eye, "unknown to fame," but no less powerful and influential.

Indeed, I will argue that it is in this domain—the private, the domestic, the familial—where wisdom has its greatest lifelong impact. Gandhi was shaped by his mother's saintliness, Benjamin Franklin by his father's practical sagacity; Confucius's diligence was rooted in a childhood with a single parent, and perhaps Socrates' hard-nosed form of philosophy owes something to his having had a stonecutter for a father and midwife for a mother. All of them, moreover, had teachers and mentors. Wisdom is apparent in the pronouncements of great leaders at moments of great historical challenge, but also in the quotidian reassurances and bits of advice shared by parents and children (a kind of wisdom that travels well in both directions, I might add). We can find it at home, on the job, in solitude and amid a crowd, in places of worship, and sometimes even in the locker room (sportswriters have long appreciated the fact that there's often more wisdom among losers than among winners).

So wisdom occupies many different venues, depending on the historical period, the cultural circumstances, and the nature of the personal or social dilemma being confronted, and is shaped by the temperaments of the people who are wrestling with those dilemmas. In an age of reason, thought will seem like wisdom's most esteemed companion. In an age of sentiment, emotion will seem like the wisest guide. If we want to push the envelope a little, in a period where basic survival is paramount, like the times we find ourselves in right now, a very practical form of wisdom is likelier to lead to a good life (in fact, a crude form of social practicality probably passed for proto-wisdom in prehistoric times). And in an age of science, the inner workings of the human brain may appear to offer us a glimpse at the biology of wisdom.

On a January weekend in 2008, an unusually diverse interdisciplinary group of researchers came together for a meeting at New York University. Some were economists, and some were ethologists, the scientists who study behavior in animals (usually primates). Some were psychologists, and some were neuroscientists. The occasion was the ninth biennial symposium on "Neuroeconomics: Decision Making and the Brain." As countless popular science books have already recounted, neuroeconomics is the relatively new but rapidly evolving field that has pushed human economic decision making into every nook and cranny of the brain. Not, it should hastily be added, without some problems still to be worked out. As one expert in "behavioral finances" put it, "Decision theory works very well in controlled situations, but works very bad in the real world, and humans operate very well in the real world." We would all be wise to mind that gap.

Over the course of the three-day meeting, there were enough statistics to trigger math-inflected migraines, and enough brain scans to put you in mind of digital phrenology. There were elaborate formulas, obscure as hieroglyphics, that purported to capture human behavior, and multiple sightings of the vaunted "hyperbolic discount curve," a simple graph that aspires to explain human impulsiveness, human impatience, and indeed human foolishness—why we always succumb to actions that sabotage our nobler long-term goals, even though we know we're being dumb.

But every talk, no matter how technical, was at least dusted with a surprisingly familiar vocabulary, a common verbal currency that would be recognizable to anyone from Socrates to your next-door neighbor: patience, delayed rewards, deliberation, reflection, decision making, attention, altruism, punishment, the role of emotion in driving desires, the role of thought in curbing those desires. As I listened to many of these talks, and struggled to understand their implications, I found myself suddenly thinking, These guys (and, unfortunately, it was mostly male scientists who spoke) are talking about wisdom, and they don't even know it.

The world doesn't need another book about neuroeconomics, and this doesn't plan on being one. But a lot of recent research in neuroeconomics and (in a broader sense) social neuroscience—including related fields like cognitive neuroscience, behavioral psychology, moral philosophy, and the like—strikes me as an immensely fertile area to till for fresh new insights into the nature of wisdom. The most successful strategy for the advancement of biological knowledge in the past half century has been reductionism—breaking down a scientific problem, or natural mystery, into its smaller, component parts and then designing experiments to tease apart the underlying biology. Using this strategy, bacteria can tell us how our genes work; fruit flies can tell us how memory works; mice can tell us how stem cells work; and now college undergraduates, who have become the model organism for much of social neuroscience, are telling us (up to a point) how the brain works.

In a metaphoric sense, I'm taking the same reductionist approach to wisdom. I've tried to break this very large idea down into several of its most salient cognitive and emotional components—I think of them as "neural pillars of wisdom," to which the second section of this book is devoted—and then paid visits to scientists doing research in those areas. This approach is wholly speculative, deeply unauthorized (at least by the scientists whose work I'll describe in the context of wisdom), and yet constantly edifying. If you ask scientists about the "science of wisdom," you'll get blank looks and rolled eyes. But if you ask about a specific, more "reductive" aspect of wisdom—emotional regulation, say, or delayed gratification or moral choice—suddenly there's a lot to talk about, think about, and, often, argue about.

Now, reductionism also happens to be, in my opinion, the most frustrating strategy for the advancement of biological understanding. When we pare and trim the big idea down to its component parts, we never quite know if we've made fatal simplifications (a point some neuroscientists concede) and we never quite know what we're throwing away. (What used to be dismissed as "junk DNA" is now seen to be a chromosomal closet crammed full of genetic control elements and evolutionary runes.) In reducing wisdom to some of its salient qualities, I plead guilty in advance to losing sight occasionally of the rich, ineffable, holistic essence of the idea itself. The problem with reductionism is that, at the end of the day, you need and want to put all the parts back together. I can't do that with wisdom; no one can. The best I can do is to respect its essential mystery while offering a peek at some of its neural gears.

And, yes, a number of scientists and organizations have begun talking about the neural components of wisdom. In 2008, the Center for Cognitive and Social Neuroscience at the University of Chicago launched a $2 million research program called "Defining Wisdom" and, in conjunction with the John Templeton Foundation, invited young neuroscientists, historians, theologians, and other academic researchers to submit grant proposals; the project later awarded twenty-three competitive grants to investigate aspects of wisdom. Even a tough-minded, hard-core cognitive neuroscientist like Stephen M. Kosslyn, who heads the Department of Psychology at Harvard University, impressed me with his willingness to engage the idea of brain function and wisdom. "Wisdom presumably has something to do with memory and reasoning, and our understanding of both has changed dramatically in recent years," he told me. "Memory is not just one thing, but rather there are many different kinds of memory, and some forms of wisdom probably rely on types of memory we didn't even know about before. In reasoning, we now know that emotion plays a major role in how we reason, and wisdom may have a lot to do with knowing when emotion is helpful and when it is not." Kosslyn also mentioned a relatively recent concept in cognitive psychology known as "framing," which refers to the way we conceptualize a problem. "People who are wise can interrupt, take a step back, and reframe," he said, "and a lot of wisdom probably has to do with looking at a situation differently and reframing."

Let me sketch out a few general principles that often seem to be associated with wise behavior. Wisdom requires an experience-based knowledge of the world (including, especially, the world of human nature). It requires mental focus, reflecting the ability to analyze and discern the most important aspects of the acquired knowledge, knowing what to use and what to discard, almost on a case-by-case basis (put another way, it requires knowing when to follow rules but also when the usual rules no longer apply). It requires mediating, refereeing, between the frequently conflicting inputs of emotion and reason, of narrow self-interest and broader social interest, of instant rewards or future gains. Moreover, it expresses itself through an insistently social vocabulary of interactive behavior: a fundamental sense of justice (which is sometimes described as an innate form of morality, of knowing right from wrong), a commitment to the welfare of social (and, for that matter, genetic) units that extend beyond the self, and an ability to defer immediate gratification in order to achieve the greatest amount of good for the greatest number of people.

Beg to differ? Good. As this brief and inadequate first pass at a definition suggests, one of the most appealing things about wisdom is the elevated form of self-awareness it inspires. When I consider the importance of emotional evenhandedness, I think immediately (as most parents would) about daily interactions with my children. When I consider "socioemotional selectivity theory," which describes how a person's emotional priorities change with shrinking time horizons (due to age, illness, or unsettling external events like the September 11 tragedy), I can't help but think about my own station in life: a baby boomer in his mid-fifties, with aging and ailing parents on one side and young, impressionable children on the other.

To repeat: Thinking about wisdom almost inevitably inspires you to think about yourself and your relationship with the larger world. With diligence (and luck), it might even make you think about how both can be made better.

Could there be a "science" of wisdom? And if there is, can it provide us anything more at this point than a fuzzy geography of neural activity superimposed upon a vague definition of a human virtue? Can it shed light on the process by which each of us deals with the decisions and dilemmas of our own private 9/11s? Can it guide us to make the best decisions possible for our loved ones and ourselves, and help us find the right path when those interests collide? Might it even hint at ways we could train our hearts and minds to give us a better shot at achieving that lofty goal?

As I embarked on this investigation, I was immensely grateful for the opportunity to find out, but terrified that I had set off on a fool's errand. As Peter Medawar, the British immunologist and Nobel laureate, once put it, science represents that rare balance of imagination and critical thinking that yields "rectifying" episodes that tell us whether a story that sounds good also rises to the level of *truth*. The story of wisdom has always sounded good, but is there anything that rectifies the notion that it has a particular biology, a scientific reality, a natural history?

Paul Baltes, in a wry bit of understatement, once described wisdom as "a topic at the interface between several disciplines: philosophy, sociology, theology, psychology, political science, and literature, to name a few." Standing at the crossroads of all those disciplines, I found it hard to know where to begin. Although science journalists feel most comfortable writing about science, that didn't seem like the best place to start. Rather, I wanted to start with what might be considered the mother of all midwives to science: philosophy.

It would be a stretch to say that Socrates or the Buddha designed the protocols for contemporary experiments in social neuroscience, but it is no stretch at all to say that a lot of the most exciting modern experimentation is founded on an empirical vocabulary that has been defined, revised, debated, contested, squelched, and resurrected over the past 2,500 years. This vocabulary of timeless human virtue—patience, moral judgment, compassion, emotional self-control, altruism, and so on—forms the foundation for what I call the eight neural pillars of wisdom, the science of which is discussed later in the book.

But before we get to questions about wisdom and the brain, we need some provisions for the trip. We need to know, at least tentatively, what was originally meant by wisdom. And then we need to pay brief visits to some of the pioneering researchers on the subject—to the philosophers who invented wisdom, and to the psychologists who first invented a way to study it empirically.

CHAPTER TWO

THE WISEST MAN IN THE WORLD

The Philosophical Roots of Wisdom

Applicants for wisdom do what I have done: inquire within.

—Heraclitus, *Fragments*

ONE DAY, late in the fifth century B.C., a well-known man-about-Athens named Chaerephon made his way to the Greek city of Delphi and posed an unusual question to the renowned oracle of the ancient world. From the time they were boys, Chaerephon had been a friend and, later, a disciple of the philosopher Socrates; indeed, both had been lampooned by Aristophanes in *The Clouds* as philosophic charlatans, and they shared many ideas about dialogue, disputation, and the steadfast, often impolitic pursuit of truth. Was anyone, Chaerephon asked the oracle, wiser than his old childhood friend? The oracle replied that no one exceeded the wisdom of Socrates.

During his famous trial in 399 B.C., Socrates recounted this story to a jury of fellow Athenians when he attempted to defend himself against attacks on his reputation—in fact, he facetiously called the oracle of Delphi as a defense witness. Although he was formally charged with corrupting young people and refusing to believe in the Athenian gods, nothing less than Socrates' lifelong pursuit of wisdom itself was on trial. Yet his greatest crime—or, perhaps, his greatest lapse in social judgment—may well have been the deft, dispassionate inquisitions by which he established that so many of his judges and jurors were not nearly as wise as they thought. As a nearby hourglass drained perhaps the final moments of his freedom, Socrates conceded that the anecdote about Delphi had contributed to "this false notoriety" about his sagacity, but he went on to admit, "I have gained this reputation, gentlemen, from nothing more or less than a kind of wisdom. What kind of wisdom do I mean? Human wisdom, I suppose. It seems that I really am wise in this limited sense."

Human wisdom? Is there any other kind?

Well, yes. There is divine wisdom, of the sort that prevailed during the centuries before Socrates stepped up to the bar, and which roared back to prominence in the Middle Ages. There is practical wisdom, of the sort immortalized in sayings and proverbs. There is even state-sponsored wisdom, of the sort that Plato would be among the first (but surely not the last) to inflict on fellow citizens. In a two-word phrase—so deceptively simple that, if you are like me, you probably didn't even notice it at first reading—Socrates managed to say something monumental in the history of human thought: that wisdom is a *human* virtue, won like all virtues by hard work, in this case the hard work of experience, error, intuition, detachment, and, above all, critical thinking. It is counterintuitive, adversarial, unsentimental, demythologizing, anything but conventional. Most important, Socrates' wisdom is *secular*, perhaps the highest form of human excellence any mortal can hope to achieve without the help of the gods (or God).

The oracle's puzzling declaration inspired Socrates to undertake what he called a "cycle of labors" to understand what exactly the god of Delphi had intended. "I am only too conscious that I have no claim to wisdom, great or small," the philosopher told his accusers; "so what can he mean by asserting that I am the wisest man in the world?" Incited by the mystery of that oracular pronouncement, Socrates then embarked upon a kind of philosophic road show (one that has as much relevance to our modern concepts of wisdom as it did to his).

On his impromptu "wisdom tour," Socrates managed to alienate, humiliate, illuminate, and educate his countrymen about the paradoxes of Socratic wisdom. He first visited a prominent Athenian politician— "a man with a high reputation for wisdom," he told the court—and lured him into a classic Socratic sand trap. The "slow old man," as Socrates described himself, wielded his questions like a box cutter, methodically stripping away ill logic and so enlarging the holes in the politician's threadbare arguments that he left the man denuded of both premise and dignity. This exercise left the philosopher disillusioned ("although in many people's opinion, and especially in his own, he appeared to be wise, in fact he was not"). When Socrates had the temerity to point this out, he managed only to provoke precisely the kind of resentment, in both the man and in his influential friends, that had led to his indictment. As he walked away from the encounter, Socrates concluded that the politician "thinks that he knows something which he does not know, whereas I am quite conscious of my ignorance. At any rate it seems that I am wiser than he is to this small extent, that I do not think that I know what I do not know." With this remark, Socrates defined an essential and indeed profound aspect of true wisdom: recognizing the limits of one's own knowledge.

The same pattern ensued with each subsequent conversation. When he moved on to the poets, Socrates felt certain that in the company of such rarefied talents, he, the poor and frankly ugly son of a stonemason and a midwife, would be exposed as a "comparative ignoramus." Again, those with the most vaunted literary reputations wilted under close questioning, so much so that a random bystander, Socrates concluded, "could have explained those poems better than their actual authors." When the philosopher turned his attention to the skilled craftsmen of his city, he searched for wisdom in what we might consider a working-class, seat-of-the-pants practical knowledge. Their technical expertise far exceeded his knowledge, Socrates conceded, and amounted to a kind of wisdom. But these skilled workers succumbed to the same affectation as the poets and politicians; "on the strength of their technical proficiency," Socrates realized, "they claimed a perfect understanding of every other subject, however important; and I felt that this error eclipsed their positive wisdom." Again, the arrogance of limited expertise inevitably produced foolishness.

At the end of his search for wisdom, Socrates decided to interrogate himself. "I made myself spokesman for the oracle," he told the Athenian court, "and asked myself whether I would rather be as I was—neither wise with their wisdom nor ignorant with their ignorance—or possess both qualities as they did. I replied through myself to the oracle that it was best for me to be as I was." And that, doubtful though its feigned humility may be, places wisdom close to its headwaters: the kind of self-awareness that allows humans to have virtues in the first place.

· · ·

Countless books and endless commentary have been written about the trial of Socrates, yet one of its most astonishing features often goes unremarked. When was the last time a "trial of the century," in any century, devoted so much public discussion to the meaning of wisdom? When was the last time a national conversation about the qualities of this elusive virtue became Topic A of its time and culture? When was the last time an entire society found itself debating the definition and importance of wisdom as if it were a matter of life and death (as it indeed was for Socrates)?

Socrates gave flesh—abundant flesh, for he did not cut a slender figure—to our enduring cultural image of the wise man. Old (he was seventy at the time of the trial), stocky, potbellied, and balding, with a pug nose well suited to his pugnacious interrogatory style and deep-set eyes as unsentimental as those of a raptor, he emerges in the pages of Plato's *Apology* (he left not a jot of writing of his own) as prickly, brilliant, unrepentant, boastful, condescending, at times almost sadistic in the glee with which he tweaks his main accuser, Meletus. In other words, the "wisest man in the world" was no laid-back Yoda. Reading between the lines, the archetype for all our cultural notions of the Socratic wise man comes across as a pain in the ass with, by his own definition, a "stubborn perversity."

These seamy details, these gossipy and unflattering physical descriptions hint at one of the most fundamental paradoxes of wisdom: It is rooted in character, personal history, and the experience of human nature, yet it is bigger than any one individual. It exists as both edifice and fog, is both immortal and yet fleeting, in some broad human consensus about what constitutes a well-lived life. Still, some individuals are bigger than others in the scheme of wise things, and some people have uniquely spoken to our urge to find wisdom in human behavior.

A couple of years ago, a friend of mine gave me an absolutely terrific (and, of course, out-of-print) little book by the philosopher Karl Jaspers, which was entitled simply *Socrates, Buddha, Confucius, Jesus: The Paradigmatic Individuals.* Jaspers famously coined the phrase "Axial Age" to define that extraordinary historical moment, roughly between 800 B.C. and 200 B.C., when civilizations in the East and West gravitated around figures that represented new modes of thought and uniquely human paths to wisdom: Socrates in ancient Greece, of course, but also Confucius in feudal China and the Buddha in the Asian subcontinent (Jesus obviously arrived on the scene a little later). If, as Socrates suggested, wisdom is essentially a human virtue, then it is virtually impossible to separate the biographical details of a life from the philosophy that exemplified that life. That is exactly what Jaspers argued in his book, which inspired me to treat the history of wisdom as a series of visitations—sometimes, I confess, a first-time visit in my case.

The arguments Socrates made at his trial and in his later deathbed ruminations (recounted in Plato's *Phaedo*) mark philosophy's grand, almost melodramatic annexation of the study of wisdom, in all its gnarly contradictions and insoluble paradoxes. Many of the essential elements we associate with wisdom were introduced as "evidence" at Socrates' trial: the importance of humility, especially in acknowledging the limits of one's knowledge and expertise; the importance of persistent, discomfiting critical thinking and discernment (usually in the form of conversational questions) to unearth the truth; the importance of identifying and pursuing goodness; and, often underappreciated, the acceptance that true wisdom at some level is often an act of hostility against society. "No man on earth who conscientiously opposes either you or any other organized democracy," Socrates told the court, "and flatly prevents a great many wrongs and illegalities from taking place in the state to which he belongs, can possibly escape with his life. The true champion of justice, if he intends to survive even for a short time, must necessarily confine himself to private life and leave politics alone."

This Socratic ideal of the closely examined, well-lived life placed cultural markers on the definition of wisdom that would not be redeemed, either by psychology or science, for millennia. But Socrates was not alone. In roughly the same remarkable period of ancient history, in every other significant corner of the civilized world, different but equally venerable schools of philosophy, each with its own definition of wisdom, began to flower. Confucius, a midlevel bureaucrat and the son of a minor government official in the Chinese kingdom of Lu, began to formulate the foundations for a very practical syllabus of public behavior that survives as a sociobureaucratic guide to wisdom. A young man of privilege and good breeding, Siddhartha Gautama, renounced worldly possessions and began teaching the sutras that would define Buddhism.

As in the famous story of the blind men and the elephant, Socrates, Confucius, and the Buddha were grabbing and describing different parts of the same unwieldy beast, wisdom; unlike the blind men, it seems to me, they were so astute, so intuitive and supple of thought, that we can see outlines of the same animal emerging in their very separate and distinct philosophies, a point Jaspers made at the conclusion of his short book. "Their concern," he wrote, "was not mere knowledge, but a transformation in men's thinking and inward action."

Although we will never be able to peer inside the incredible minds of Socrates and Buddha with the help of an MRI machine, many of the qualities they identified as central to their conception of wisdom can, in fact, be investigated in scanning machines. It's not the pictures, pretty as they are, that interest us; it's the vocabulary of those qualities, first coined by these paradigmatic individuals and now written into the protocols for twenty-first-century experiments, that carries the remote but distinct promise of showing how our minds work as they rise to this exemplary level of human virtue. This science even provides a tentative, early glimpse of the kinds of practices and habits that might improve the quality of our minds as we, too, seek wisdom.

The stories of Socrates, Buddha, Confucius, and their contemporaries conceal important, and somewhat sobering, lessons about the cultural context in which wisdom develops and flourishes, or not. The Buddha, whose penniless wandering has inspired the enlightenment of millions, was yet not wise enough to designate a successor—an oversight that helped trigger centuries of schism and dispute in Buddhism. For the last ten years of his life, Confucius traveled widely in China in search of a job, and yet no one would hire him; when he died, historians of philosophy tell us, he felt his life had been a failure. And although the philosophy of Socrates (as filtered through the writings of Plato) has served as a template for assessing wisdom, the trial of Socrates reminds us that, even in his own time and among the hundreds of Athenian peers serving on his jury, wisdom could inspire intense social and political backlash. "Because Socrates alone had seriously digested the precept of his god— to know himself—and because by that study he had come to despise himself, he alone was deemed worthy of the name wise," Montaigne wrote. "Whoever knows himself thus, let him boldly make himself known by his own mouth."

. . .

The quest for wisdom is a physical as well as an intellectual undertaking. It often requires changes in scenery, thrives on commerce (which often promotes the exchange of ideas), and usually involves a journey. Whether it is Socrates making the rounds of Athens to interrogate politicians and poets, or the Buddha wandering through deer parks in northeast India to spread word of his awakening, or Confucius pounding the pavement in the twilight of his life in search of employment, the early history of wisdom unfolded on the road.

Some historians of philosophy argue that Homer (in the *Odyssey,* the ultimate road book) and Hesiod (in the *Theogony*) began this initial move toward a profound and fundamental "mutation" in thought as early as the ninth century B.C. with a simple, but telling, narrative omission. These epic poems do not embrace magic, and the verbs in their stories connect for the most part with human subjects, not divine acts. "Homer and Hesiod," Wallace Matson has written in his sprightly *A New History of Philosophy,* "explain only in terms of personal agency." This may sound like a modest academic observation, but such independence of thought marks a momentous shift in the cultural evolution of wisdom.

Humankind's revolutionary emancipation from magical agency and capricious gods, like all emancipations, is a blessing and a burden. We are blessed by the freedom to think and decide for ourselves, but burdened by the responsibility that comes with human actions and their human consequences. If we humans control our own activities, it naturally follows that we are responsible for our own behavior, and, just as important, responsible for sanctioning others for their *misbehavior.* This notion of social justice—and injustice—represents nothing less than an initial attempt to institutionalize wisdom, to create a political and legal structure for the discernment of right and wrong, fair and unfair, selfish and selfless, one that probably originated in the prehistoric apportionment of food, clothing, and relative proximity to a warm hearth in primitive rock shelters. Thus, it is no accident that when twentieth-century psychologists began to ponder the nature of wisdom, they paid special attention to accounts of ancient kings and judges in the Hebrew Bible and early folk texts. As Gibbon pointed out, most Roman magistrates were philosophers, so there persists into the Christian era this overlap of wisdom, human judgment, and social justice.

Finally, there is Heraclitus, who in the bright, shiny, poetic shards of philosophy that survive in his *Fragments* made contingency and change the essential challenge of wisdom. To Heraclitus, knowledge was fluid (his metaphor was fire). Nothing was constant; everything changed. And Heraclitean wisdom began with the fact that reality is dynamic; the world as we thought we knew it in the past is sure to be different as we venture into the future. It changes; we change. And so, in his most famous metaphor, when we dip our toe into a river, the river itself is different from the way it was just a moment ago, when our toe entered.

> *The river*
> *where you set*
> *your foot just now*
> *is gone—*
> *those waters giving way to this,*
> *now this.*

During a rare period of relative peace, after the defeat of Persia in 479 B.C. and prior to the start of the Peloponnesian Wars in 432 B.C., Athens enjoyed a pacific explosion of commerce, public building, the arts, and, of course, the practice of philosophy at the highest level. The city remade itself as an urban shrine to Wisdom, deified in the Parthenon, the great temple erected to honor Athena, the patroness of the city and goddess of wisdom. Philosophy assumed a central role in civic and cultural life; as Socrates told the jury at his trial, Athenians "belong to a city which is the greatest and most famous in the world for its wisdom and strength."

In a famous speech honoring soldiers who had died in the first skirmishes of the war between Athens and Sparta, Pericles spoke at length of the greatness of Athens at its peak in the fifth century B.C., and his remarks manage to capture a working definition of civic wisdom as it existed at the height of Greek civilization 2,500 years ago. He spoke of a uniquely Athenian talent for deliberation and decision making that, whether in the public domain in which he operated or in a more domestic setting, established a model for wisdom in action, one that probably strikes us as bracingly modern. "We Athenians," he said, "are able to judge at all events if we cannot originate, and instead of looking on discussion as a stumbling-block in the way of action, we think it an indispensable preliminary to any wise action at all. Again, in our enterprises we present the singular spectacle of daring and deliberation, each carried to its highest point, and both united in the same persons; although usually decision is the fruit of ignorance, hesitation of reflection."

In the Periclean formulation, wisdom was deliberative, judgmental, collective, reflective, and deeply social, rooted in conversation and disputation, steered by critical thinking. All these elements, in varying proportion, form a reasonably good working recipe for the kind of decision making that, even today, we likely would consider wise. They also begin to provide a vocabulary, a kind of high-minded philosophic slang, for modern experimentation on human judgment and decision making.

For all its celebrated enlightenment, Athens was not so enlightened a place that its citizens could resist putting the wisest man in the world to death. Plato used Socrates' last moments on earth (in his dialogue *Phaedo*) as an occasion to reiterate his belief that the pursuit of wisdom was the highest human calling—and, perhaps, a mission more easily accomplished without the distractions of bodily desires.

In a memorable passage, Socrates lamented that "the body fills us with loves and desires and fears and all sorts of fancies and a great deal of nonsense, with the result that we literally never get an opportunity to think at all about anything. . . . That is why, on all these accounts, we have so little time for philosophy. Worst of all, if we do obtain any leisure from the body's claims and turn to some line of inquiry, the body intrudes once more into our investigations, interrupting, disturbing, distracting, and preventing us from getting a glimpse of the truth. We are in fact convinced that if we are ever to have pure knowledge of anything, we must get rid of the body and contemplate things in isolation with the soul in isolation. It's likely, to judge from our argument, that the wisdom which we desire and upon which we profess to have set our hearts will be attainable only when we are dead, and not in our lifetime."

It is hard to read these lines now without recognizing a fatal, centuries-long flaw in the psychological logic of this valedictory. In Plato's account, Socrates essentially demonized and exiled emotion as a foe of wisdom, devaluing its importance, casting it as distraction and perturbation, setting it up for a philosophical reclamation project that would last more than a millennium. It is also difficult to read these lines now without thinking of neuroscience's recent notion of "embodiedness," the idea that bodily sensations do not so much compete with the mind as become satellite outposts of the mind, not only informing it but part of it.

In any event, soon after these remarks, a jailer brought Socrates the cocktail of hemlock, and he calmly drank his poison. Surrounded by disciples, the philosopher-king of the interrogatory posed his final questions: "Will a true lover of wisdom who has firmly grasped this same conviction—that he will never attain to wisdom worthy of the name elsewhere but in the next world—will he be grieved at dying? Will he not be glad to make that journey?" Even in death, Socrates managed to make wisdom a road trip.

Although standard histories locate the nativity of philosophy in Greece in the sixth century B.C., the dawn of the Axial Age rose first in the East. At least a century earlier than the Greek stirrings in Miletus, Eastern religions and belief systems had begun to coalesce around a somewhat different notion of knowledge, one that nonetheless anticipated contemporary concepts of wisdom. The Upanishads—formally committed to written form only around 800–500 B.C., roughly the same time frame of Homer and Hesiod—assembled poems and tales that offered the collective wisdom of saints and sages on a less material, almost ineffable plane of knowledge. "Wisdom diverged from the knowable, sensory world we live in," James Birren and Cheryl Svensson observed, "to a vaster, more intuitive understanding of the nature of life and death." This more intuitive form of wisdom especially flowered in the teachings of Confucius, the most influential Chinese philosopher of all time, and, later, in the Buddha.

Born in the sixth century B.C., Confucius lived most of his adult life in the north-central coastal region of China, although he traveled extensively later in life—in a ceaseless and largely unsuccessful search for work! His father, a minor military figure, was seventy years old when Confucius was born, and he died when his son was three. Confucius applied himself avidly to his studies as a young man, married at age nineteen, and had a son named Kong Li (Li meaning "carp"). Always poor, but always diligent and ambitious, Confucius studied intensely while working as a clerk and a zookeeper. As he famously remarked toward the end of his life, "When I was thirty I began my life; at forty I was self-assured; at fifty I understood my place in the vast scheme of things; at sixty I learned to give up arguing; and now at seventy I can do whatever I like without disrupting my life."

Very early on, Chinese wisdom acquired a practical, discreet, commonsensical, almost bureaucratic bent as people sought stability and pacifism. During Confucius's lifetime, the centuries-long reign of the Zhou dynasty in the western territories had collapsed, leading to great political instability, and with this came a period of intense social uncertainty and military strife. As civil order unraveled in this indifferent, feudal era, Confucius witnessed human misery on an epic scale. "This grounding in everyday horrors had a profound effect upon young Confucius," British author Paul Strathern has written. "It was to give a toughness and practicality to his thinking, which it seldom lost." This tough, unapologetic practicality is a reminder that our definitions of wisdom are shaped—and, perhaps, constrained—by unique local conditions of history and culture. And yet it is due to the power of wisdom that Confucian insights gestated in feudal China resonate instantly and effortlessly in twenty-first-century lives.

In numerous parts of the *Analects,* his collected sayings, Confucius reiterates a triad of foundational principles that he believes should guide the good life. Most important is goodness (or *gen*), which has become, arguably, one of the most powerful beacons illuminating human behavior in the history of civilization. In the Confucian hierarchy of virtue, even wisdom and courage were secondary to *gen.*

But Confucius's enduring greatness and influence derive from his understanding that in order to reduce the vast human misery he saw all around him, society needed to change, and in particular, it needed to reorganize its goals in order to seek the benefit of the many rather than the comfort of the few. He understood that this transformation might best be accomplished by the politic (in every sense of the word) and practical behavior of, in effect, civil servants. Hence, Confucianism often reads like a bureaucratic catechism about civic altruism or public comportment—or, more specifically, virtue or morality as it unfolds in the public arena.

Many elements of Confucianism anticipate subsequent developments in Western philosophy—for example, the commitment to social justice in Greek society ("Not to act when justice commands, is cowardice," Confucius wrote), and a moralism grounded in public virtue, with obvious precursors to Christian loving-kindness. "For Confucius, morality was all about *involvement* in society," Strathern writes. Confucius also emphasized the primacy of emotion (as it was expressed in compassion and intuition) over reason, and the melding of personal moral behavior with the larger political order. This is one of the reasons that Confucian teaching still offers some of the shrewdest advice about wise behavior in a government bureaucracy or the corporate workplace.

Despite his enduring influence, Confucius, like Socrates, did not enjoy an abundance of loving-kindness from his contemporaries. The prince of Lu appointed the middle-aged Confucius to be the minister of crime, but his term in office degenerated into something of a reign of terror. He was said to have been hugely successful at eradicating crime, but his zeal sometimes exceeded his wisdom—purportedly, he condemned to death, for example, anyone guilty of inventing "unusual clothing." This ill-fated imposition of wisdom recalls Plato's equally zealous proposals to ban Homer, drama, and slow music for children. But encountering these bouts of bewildering silliness, it occurred to me that even the wisest people the world has ever known can do terribly unwise things, especially when they attempt to impose their version of wisdom upon others. Those who aspire to wisdom must be drawn to it, and seek it, not receive it as a government regulation, spiritual proclamation, or philosophic incitement to acceptable and virtuous thought.

Ultimately a polarizing figure who was relieved of his duties, and unemployed in his fifties, Confucius began a ten-year pilgrimage across China in a fruitless search for a job. Although history celebrates Confucius as one of the wisest human beings ever to have lived, none of his contemporaries would hire him, and he thought of himself as a failure. He died in 479 B.C., his last words reputedly:

> *The great mountain must crumble,*
> *The strong beam bursts,*
> *The wise man must wither away like a plant.*

Especially in early Eastern philosophy, an acute appreciation of human fallibility emerges as a fundamental construct of the human condition. Confucius wrote, "To know what you know and know what you don't know is the characteristic of one who knows," an almost exact precursor to the point Socrates made to the jurors at his trial nearly a century later.

This embrace of limitation, of uncertainty, becomes deeper, richer, and even more liberating in the teachings of the Buddha. When Prince Siddhartha Gautama abandoned a life of leisure and privilege shortly before the age of thirty, he famously repaired to a forest in east India, sat at the foot of a Bodhi tree on an "adamantine throne" of pressed grass, and declared, "I will not uncross my legs until the destruction of defilements has been attained." These initial vows, however, may have been shaped by adversity; Gautama's retreat to the forest followed his rejection by a group of fellow ascetics as a "backslider" whose commitment to devoted meditation was suspect. Left on his own, the prince—scorned by the others for "living luxuriously, slacking in his effort, and backsliding into luxury"—went on to experience one of the most shudderingly transformative spiritual awakenings known to the human condition.

Almost an exact contemporary of Confucius (as exact as ancient records allow) and about a century earlier than Socrates, Gautama was born around 563 B.C. and died around 483. Prince Siddhartha followed a different path to wisdom. The Buddha taught a form of enlightenment that emphasized knowing through personal experience, suppression of sensory urges, and a cultivation of selflessness. In their reading of the famous Sutra of the Four Noble Truths, psychologists Birren and Svensson conclude, "Wisdom lies in stilling all desire." As Jaspers put it, the Buddha "strives to annul the world by extinguishing the will to exist."

Although Buddhism embraces "awakening," not wisdom per se, the spiritual rewards that accompany this awakening translate well into the art of living that we have come to associate with wisdom: equanimity in dealing with the unknowability of answers, mastery over desire, the dampening of material selfishness, and an ability to thrive despite uncertainty. In a larger sense, however, Buddhism shared a crucially important conceptual shift with ancient Greek philosophy. Both philosophies shed divine authority as a source of knowledge, and both celebrated human-based insights into the nature of reality. True wisdom, both in the East and the West, begins with the individual.

Having experienced his awakening under the tree, Gautama emerged with a startling statement: "Conquerors are those who, like me, have dissolved habituated desires." That thought probably sends a shiver of recognition through current practitioners of behavioral psychology, where the daily coin of the experimental realm is habit and desire. The Buddha made his way to a deer park near Varanasi, in northeastern India, where he rejoined the five mendicants who had previously spurned him. In a famous sutra called "Turning the Wheel of the Dharma," Gautama described to his colleagues a "middle way" between the "practice of clinging to sensory pleasure in sensory objects" and "the practice of exhausting oneself with austerities." This balance, he explained in another sutra, accompanied by "complete attentiveness," allows one to tune out birth, death, aging, existence, and all craving and grasping. "And there arose in me," he declared, "an eye for previously unheard of matters, there arose in me direct knowledge, penetrating insight, wisdom, and clear seeing."

"The Fortunate One" spent the next forty-five years wandering throughout east India, imperturbably homeless and happily dispensing this message; the teachings, preserved by oral tradition for centuries, were ultimately transcribed onto birch bark by monks in Sri Lanka around A.D. 17, and they remain an unparalleled (though not uncontested) repository of Eastern wisdom. Many of the sutras delineate behaviors that should be "censured by the wise," including infatuation, hostility, delusion, qualities that create a person who "takes what is not freely given, engages in damaging sexual relations, speaks falsely, and incites others to do just the same."

As the Buddha told his fellow mendicant Kalamas in "Discourse in Kesamutta," "the superlative practitioner, who, thus embracing beneficial teachings, becomes free from desire, free from hostility, without confusion, attentive, and mindful, dwells gradually pervading the world with a heart suffused with friendliness." In the modern psychological idiom (and, increasingly, in the idiom of contemporary neuroscience), the Fortunate One described cognitive and emotional qualities that could be reductively parsed and studied, including emotional regulation (lack of hostility), compassion, attention, and the elevated state of cognitive focus known as mindfulness. As we'll see in a later chapter, neuroscientists have in the last few years measured the brain activity of Buddhist monks while they meditate, and indeed their neural activity *is* different— although what exactly that difference means remains, true to the deeper message of the Buddha, somewhat uncertain.

As distinct as these schools of thought are, it is their deeper congruences that begin to coalesce around a time-tested, culturally heterogenous, geographically far-flung, yet surprisingly universal concept of wisdom. East or West, they all embrace social justice and insist on a code of public morality. They embrace an altruism that benefits the many. They try to dissociate individual needs and desires from the common good, and strive to master the emotions that urge immediate sensory gratification. And in their choice to be teachers, Confucius and Socrates and the Buddha, each in his own way, asserted the central primacy of *sharing* their accumulated body of life knowledge. That impulse would culminate in the creation, back in Greece, of the first formal academy—a school in which, it might reasonably be said, everyone majored in wisdom.

To some degree, the natural history of wisdom can be seen as a never-ending battle between the forces of theology and those of secularization, between a top-down, benevolent, dispensed, and divine form of wisdom and a bottom-up, organic, hard-earned human form of wisdom. Put simply, is wisdom a human quality, achieved by human intelligence and insight? Or is wisdom a heavenly gift, bestowed by the gods (or God), utterly inaccessible to mortals who do not subscribe to one or another of the world's religions? This question was embedded in the earliest versions of the Hebrew Bible, in Eve's apple and Solomon's dream, but it became especially thorny with the rise of Christianity, which restored an era of authority, nonhuman agency, and religious mysticism. In such a setting, Wisdom (with a capital *W*) comes from on high, and wisdom (with a small *w*) contracts to the rather more modest spheres of familial and communal enterprise.

No single person embodied theology's recapture of wisdom more than Augustine, who single-handedly demoted secular wisdom to a lower, more inferior plane in the hierarchy of human achievement. Saint Augustine distinguished between *"sapientia,"* a timeless, eternal wisdom about human conduct and moral perfection bestowed by God, and *"sci-*

entia," which reflected mere knowledge of the material world. Taking their cue from Augustine, religious authorities drew an increasingly sharp (and intolerant) distinction between the material wisdom based on reason and knowledge on the one hand and, on the other, the spiritual wisdom and knowledge associated with faith and ecclesiastical authority, a segregation marked most famously, if not first, in the story of King Solomon.

Solomon's iconic wisdom, after all, came to him in a dream, a divine gift from God; for Jews, wisdom arose out of a personal relationship with God. This important split between Science and Wisdom, between human-perceived material truths and human-*received* universal truths, dominated philosophy and religion for the better part of a thousand years, and it is a dichotomy that we're still working through. (Indeed, the John Templeton Foundation, which describes itself as "a philanthropic catalyst for discovery in areas engaging life's biggest questions," has begun to fund scientific and other scholarly research into wisdom, which in turn has led some scientists to complain that the foundation has an agenda to use science to promote and legitimize religious beliefs and values, including wisdom.)

The Augustinian notion of wisdom poses a fundamental challenge to us even today. On the one hand, there is growing evidence—although many scientists seem oblivious to the idea—that religion promotes precisely the kinds of communal values and interactions that social neuroscience is most keen to study: compassion, altruism, other-centered thinking. As a facilitator and mediator of responsible community interaction, the *culture* of religion can easily be seen as a force in cultivating the social dimension of wisdom. On the other hand, the relative inflexibility of religious doctrine is by its nature anathema to the contextual suppleness of wisdom. Which leaves us with an uncomfortable question: Does religion ultimately promote, or undermine, human aspirations to wisdom? This book won't come close to settling that nettlesome question. But Paul Baltes, the German psychologist and arguably the foremost scientific scholar of wisdom, wondered aloud late in life whether religion, because of its insistence on a fixed set of values, was an "intellectual enemy" of wisdom. While acknowledging that organized religion makes wisdom a cultural and spiritual priority, he concluded that "religions constrain how far wisdom is developed. In fact, there may be a point beyond which religion becomes a hindrance to the generalizability or transcultural validity of wisdom."

The Judeo-Christian tradition of wisdom thus leaves us with a paradox. It clearly regards wisdom as a divine gift. Yet it also promotes an understanding of human behavior, and the use of that understanding to guide (if not mold) the behavior both of individuals and, perhaps even more important in a cultural sense, of social groups ranging from immediate family to kinship groups to larger, nonconsanguineous social organizations. In other words, the religion-based strain of wisdom extended well beyond the "examined life" of Socrates and enlarged the circumference of its meaning to include family, community, ethnic group, creed group (for all his wisdom, Socrates does not come across as much of a family man). The book of Proverbs can be read as among the earliest of self-help manuals. Its mantric code of prescribed behaviors—the serial, repetitious injunctions against lust, infidelity, imprudent business affairs, and similar pragmatic advice—is packaged as a wisdom instructional; the rhetorical style of its narrative establishes one of the iconic modes of transmission for wisdom: father to son or, more generally, parent to child, elder to youth. Here, bald emotion—notably, the fear of transgressions that would invoke God's wrath—guides wise behavior.

It took many centuries (and, probably, the death of many "wise" contrarians), but Renaissance thought restored the central secular importance of wisdom, as well as its fundamental subversiveness. In a wonderful little book entitled *The Renaissance Idea of Wisdom,* the scholar Eugene F. Rice, Jr., wrote, "Augustine tied *sapientia* and Christianity together with knots which held a thousand years. The Renaissance patiently loosed them, and restored wisdom to its old autonomy and its purely human dignities."

Montaigne turned his pen into the biggest drill bit of the Middle Ages. He seized wisdom from the realm of received goods and turned it into a high-end, artisanal craft—individualized, even idiosyncratic, privately verified and ratified, and reflexively wary about disguising layers of artifice and authority. The *Essays,* to which I'll return again and again, were nothing less than the invention of the personal essay as a high-powered literary microscope, much as Galileo would later invent the telescope, an instrument used to discern, examine, and discover startling new truths about very familiar things.

The wise person of the Renaissance needed to do her or his philosophic due diligence, to scratch beneath the surface of things to truly understand them, whether it was human nature or an aspect of physical reality. If conventional wisdom (in any age) represents a kind of authority, this was the rebirth of wisdom as a subversive, antiauthoritarian force. This newfound (or, more properly, rediscovered) commitment to use questions as a digging tool encouraged a behavioral ethic of challenging thought, authority, and, of course, conventional wisdom.

Wisdom has never ceased to be a formal and central concern of philosophy. "The Idea of wisdom," Immanuel Kant wrote, "must be the foundation of philosophy. . . ." Yet this foundation, in his view, rested on shaky, if not invisible, legs. Kant conceived of two different, mutually exclusive realms of reality that by definition constrain our embrace of wisdom: the phenomenal world, "where knowledge is possible," and what he called "the noumenal," "which is transcendent and to which there is no access." Once again, we can hear the echo of Eastern philosophy rattling around Kant's lonely room in Koenigsberg, with its very conscious acknowledgment of limitation. Kant also anticipated the life-span psychologist Paul Baltes (whom we'll meet in the next chapter) in his belief that wisdom was an ideal, always to be aspired to but essentially unattainable. Humans, Kant believed, "did not possess wisdom but only felt love for it."

The French philosopher Jean-François Revel argued that philosophy essentially abandoned wisdom as a worthy goal in the eighteenth century. Prior to Descartes and Spinoza, philosophy unified the pursuit of science with the pursuit of wisdom. Spinoza, he said, represents "the final appearance of the idea that supreme knowledge can be identified with the joy of the sage who, having understood how reality works, thereby knows true happiness, the sovereign good. . . . Over the last three centuries, philosophy has abandoned its function as a source of wisdom, and has restricted itself to knowledge." Indeed, by emphasizing the importance of knowledge—and thus its attainment—philosophy became midwife to astronomy, physics, chemistry, and biology.

And yet, wisdom still retains its grip—on philosophers and on us. Why? The contemporary philosopher Robert Nozick attempted to provide an answer in a 1987 essay entitled, "What Is Wisdom, and Why Do Philosophers Love It So?" His answer was quite simple. "Wisdom," he

wrote in the very first paragraph, "is an understanding of what is important, where this understanding informs a (wise) person's thought or
action." And then, as philosophers are wont to do, he went on for
another ten pages qualifying that definition.

If the last few paragraphs seem like a whirlwind tour of philosophic
name-dropping, I apologize—up to a point. The aim is not to be drearily
comprehensive: that, in any event, would be both pointless and impossible. Rather, I'm trying to hint at, in as concise a method as "drive-by"
citation allows, the extraordinarily long and rich intellectual genealogy
for what is to come. Many of the "lovers of wisdom" whose names you
have just read have been cropping up at a surprising rate in the scientific
literature over the last decade or so, specifically in the field of neuroscience. From Antonio Damasio's immensely popular books, such as
Descartes' Error and *Looking for Spinoza* to Joshua Greene's piquant essay
"The Secret Joke of Kant's Soul" to the citations of David Hume (and
Aesop, for that matter) in recent papers that have appeared in *Science*
and *Neuron,* there has been a convergence of cutting-edge curiosity
between those traditional "lovers of wisdom" and modern scientists
about how the mind works. It is no stretch to say that many of the cognitive processes being scientifically pursued hew to this long philosophic
history of thinking about wisdom.

But before we set out to see if there is a *biology* of wisdom, I want to
invite one other field into the conversation. A decisive turn in the formal
history of wisdom research occurred in the 1950s, when the psychoana-
lyst Erik Erikson suggested that wisdom was a defining feature of one of
the last stages in human development. This seemingly modest observation—Erikson's first mention of wisdom amounted to a single word
appearing in a graph about his eight stages of life; he initially prepared
the graph in 1950 for a White House conference on childhood development—eventually had two enormous implications. One was that wisdom became a topic that could be experimentally tackled by the field of
psychology. The other was that wisdom, however provisionally, became
formally associated with a specific stage of life—namely, old age. Once
Erikson left the door to wisdom studies ever so slightly ajar, a few brave
psychologists rushed in.

CHAPTER THREE

HEART *AND* MIND

The Psychological Roots of Wisdom

Of wisdom, therefore, which all men by nature desire to know and seek with such mental application, one can know only that it is higher than all knowledge and thus unknowable, unutterable in any words, unintelligible to any intellect, unmeasurable by any measure, unlimitable by any limit . . . unaffirmable by any affirmation, undeniable by any negation, indubitable by any doubt, and no opinion can be held about it.

—Nicholas of Cusa, *De sapientia*

THE IDEA THAT WISDOM might be a legitimate topic for empirical research can trace its roots back to an apartment building on Newkirk Avenue, just off Coney Island Avenue, in Brooklyn, New York. That is where, in the 1950s, a keenly observant young girl named Vivian Clayton became fascinated by several exemplary qualities she attributed to the two central figures in her life: her father, a furrier named Simon Clayton, and her maternal grandmother, Beatrice Domb. There was something about the way these two elders conducted their lives that distinguished them from other people in Clayton's ken. Despite limited education, they possessed an uncanny ability to make good decisions; they remained calm in the midst of crises; they conveyed an almost palpable sense of emotional contentment, often in the face of considerable adversity or uncertainty. In contemplating those unique qualities, Clayton found herself, long before she went to college, contemplating the nature of wisdom.

"My father was forty-one when I was born," she later explained. "By far, he was the oldest parent among all my friends, almost the age of my friends' grandparents. He had emigrated from England but had lived through World War II and experienced the blitz and had to care for his dying mother, who was so sick that she couldn't go down into the shelters during air raids in London. She lived in the East End, where the docks were, and they were always getting bombed. So he would sit with her while the bombs were falling, and when it was over, she would say, 'Now we can have a cup of tea!' He was a very humble man, and very aware of his limitations, but he always seemed to be able to weigh things and then make decisions that were right for everyone in the family. He knew what to respond to quickly, and what you had to reflect on."

Clayton's maternal grandmother, Beatrice, was the other central role model in this precocious fascination with wisdom. "My mother would belittle my grandmother as a simple person," Clayton said. "But her simplicity I saw as a sign of deep contentment in her own life. She, who had less than a high school education, was the matriarch of this very large family."

Vivian Clayton obsessed over the differences between her mother and father, her grandmother and grandfather. She recalls pondering these differences as an adolescent at Bayside High School in Queens, as an undergraduate studying psychology at the State University of New York at Buffalo, and, more formally, as a graduate student in the early 1970s at the University of Southern California, where she worked with one of the country's leading geriatric psychologists, James E. Birren. She is generally recognized as the first psychologist to ask, in even faintly scientific terms, "What does wisdom mean, and how does age affect it?"

Although Erikson is generally credited with staking the claim that wisdom could be thought of as a stage in human development, he did not pursue the subject much after making this claim in the 1950s. Erikson, however, was uniquely positioned to see wisdom as the culmination of a lifelong process of emotional maturation. After coming to the United States from Europe in the 1930s, he had the good fortune to participate in the pioneering Berkeley Growth Study at the University of California, one of the earliest psychological surveys of its time to follow the same group of individuals as they progressed from the toddler stage to young adulthood.

Erikson, who cut his teeth, according to psychiatrist George Vaillant, "on the first great longitudinal studies of human development at Berkeley, argued convincingly that adults, like children, evolve and mature."

As Vaillant observed in his book *Adaptation to Life*, "Certainly, Shakespeare had said it all before, but most textbooks of human development associate changes in adult personality with external events." Remarkably, Erikson seems to be the first psychologist to have suggested that wisdom could be acquired through a stepwise, lifelong process of self-realization.

Erikson viewed wisdom as a central feature in what he called the "eighth stage" of psychosocial development. Certain individuals, he believed, achieve enough emotional resilience (or "ego integrity") over the course of a lifetime to overcome the despair that arises as the end of life approaches. Later, in *Identity: Youth and Crisis* (1968), Erikson elaborated—ever so slightly—on these ideas. Wisdom, he wrote, most likely arose during "meaningful old age," after nearly a full lifetime of accumulated experience but before "possible terminal senility" set in, when an individual could look back with perspective on an entire life. "Strength here," Erikson wrote, "takes the form of that detached yet active concern with life bounded by death, which we call wisdom in its many connotations from ripened 'wits' to accumulated knowledge, mature judgment, and inclusive understanding. Not that each man can evolve wisdom for himself. For most, a living 'tradition' provides the essence of it."

Several terms here are key because of their fecund psychological (and ultimately neuroscientific) implications: "detached" (with its implications for the regulation of emotion), "inclusive" (because it makes wisdom not only social but *encompassingly* social in understanding another's perspective), "tradition" (with its suggestion that ritual, culture, history, and family are all repositories of wisdom), and finally "bounded by death" (a dramatic demarcation reminding us that time and impending mortality impose a crucial temporal dynamic to any notion of wisdom). By invoking wisdom as perhaps the highest form of human psychosocial development in several books in the 1950s and 1960s, Erikson opened the door to its formal study by social scientists, with Vivian Clayton taking the lead. Erikson never scientifically defined wisdom, however, and in order to study a psychological phenomenon, psychologists need to "operationalize" the concept—figure out a way to study it quantitatively, beginning with a definition of what they propose to measure.

Clayton's modern consideration of wisdom began with a tremendous bias, but a good one, which counterbalanced an equally powerful bias that had soured the biomedical literature on aging in the 1960s and

1970s. Half a century ago, gerontologists dominated the study of old people, typically conducting studies on elderly patients in hospitals and assisted-living facilities; although only 5 percent of the elderly lived in nursing homes, according to researchers, almost all the gerontological research focused on this frail and struggling institutionalized population. Like the proverbial drunk looking for his car keys under a streetlamp, these researchers looked for, and failed to find, anything positive to say about growing old.

One of the leading voices pushing for a broader, more balanced, and more global view of the aging process was Birren, a pioneering psychologist at the Ethel Percy Andrus Gerontology Center at USC. At the same time Clayton began her doctoral studies, Birren encouraged other of his graduate students to examine messy but important aspects of aging—not just wisdom but things like love and creativity. In what amounted to a battle between psychologists and gerontologists for the soul of aging research, Birren made a commitment to investigate positive aspects of aging, including wisdom, and was soon joined by another prominent social scientist: Paul Baltes. Then a psychologist at Pennsylvania State University, Baltes helped pioneer life-span development theory, an influential school of psychological thought that argues that in order to understand, say, the psychological state of a sixty-year-old, you need to take into account the person's biology, psychology, and social context, as well as the cultural and historical events through which she or he has lived. Life-span studies placed a new and sophisticated emphasis on the psychological significance—and this would be important to later conceptions of human choice, including brain-scanning experiments—of time in shaping people's behavior. It also promised a methodological way to focus on that sliver of life span that Erikson had identified as key to wisdom: after a lifetime of maturation and experience, but before senility.

In the early 1970s, when Vivian Clayton began to think about the topic, there was no "wisdom literature" in psychology. Birren sent her off to consult ancient accounts of wisdom, which introduced her to what is, in many ways, the most frustrating and yet fascinating aspect of the field: How in the world do you even define it?

Clayton flung herself into precisely the same nonscientific literature that represents the repository of human thought about wisdom: Eastern religion, Greek philosophy, and, perhaps most interesting to her, the venerable "wisdom literature" of the Old Testament and related parables of wisdom from the Hebrew tradition. By the time she had finished with the five books of Moses and the classic "wisdom" books of the Bible (Proverbs, Job, Ecclesiastes, Ecclesiasticus, and Song of Solomon), she had reached a kind of epiphany, one that has reverberated throughout psychology and, more recently, neuroscience.

In Clayton's view, wisdom was different from intellect and necessarily went beyond mere cognitive ability. While intelligence, she wrote, could be defined as the ability "to think logically, to conceptualize, and to abstract from reality," wisdom extended knowledge to the understanding of human nature, of oneself as well as others, and yet operated on "the principles of contradiction, paradox, and change." In short, intelligence represented a kind of knowledge that was fixed, impersonal, and, in an odd sense, nonsocial. Wisdom represented knowledge that was, by contrast, profoundly social, deeply personal, adaptive, and intuitive. It incorporated (to use a term of more recent coinage) *emotional* intelligence. This was the essence of the ideas that Clayton, more or less single-handedly, began to lay out in her early work in the field.

The famous story about the wisdom of King Solomon (in 1 Kings)—where he resolved the dispute between two women, both claiming to be the mother of a baby, by threatening to cut the child in half—satisfies us on many levels. Solomon's decision was shrewd; it revealed penetrating insight about human nature (only the real mother would recoil at imminent harm to her child); it averted a moral wrong; and it resolved a nettlesome problem. We all know and appreciate those dimensions of the story. But few of us acknowledge the limitations of Solomon's wisdom, and what that tells us about the psychology of this elusive virtue.

To begin with, Solomon's wisdom was a gift, received from God in a dream. Moreover, his story reminds us that a wise action reflects only an imperfect correlation with the character of the individual. As we'll explore in greater detail later, Solomon's decisions and actions during his rise to power recall not wisdom, but calculation, ruthlessness, vengeance, and, above all, vanity. Once he assumed the throne vacated by his father, David, he devoted an unseemly amount of time to interior decorating, describing with almost gloating attention to detail the construction of his temple and palace. By the twilight of his life, his self-absorption and indulgences are seen by many scholars as the antithesis of wisdom. There is a moral in Solomon's life-span development: True wisdom is more complicated and elusive than any parable or morality tale can convey. And, like more material forms of wealth, the fact you now have it doesn't mean you get to keep it.

Nonetheless, Hebrew narratives, which veered off in a different direction from Greek rational thought, prove to be an exceptionally rich source for thinking about wisdom. From the very first episodes of Genesis, the distinction Clayton had begun to sketch out—between intelligence and wisdom—has parallels in the earliest events in the world. God not only created the world with wisdom; in a sense, he endowed Adam and Eve with the possibility of *human* wisdom when he expelled them from the Garden of Eden. In his recent translation of *The Five Books of Moses,* Robert Alter describes Eve's curiosity, just before she plucked and ate fruit from the tree of knowledge, as "lust to the eyes and the tree was lovely to look at."

But, as Alter points out in a note on the translation, the Hebrew verb for "to look at," *lehaskil,* can also be translated as "to make one wise"— a wonderful conflation of sensory perception and cognitive insight. So when the serpent told Eve, "For God knows that on the day you eat of it your eyes will be opened and you will become as gods knowing good and evil," perhaps there is a possible meaning different from the customary message about shame and exile. What is culturally and traditionally assumed to be a loss of innocence might also be a myth about the acquisition of wisdom—indeed, about the emotional and intellectual price of seeing things clearly and becoming wise. Instead of original sin, this version of "original wisdom" suggests not only a dispassionate gathering of knowledge ("eyes wide open") but also the gift of discernment ("knowing good and evil"), two qualities that are essential "to make one wise." This kind of wisdom, a persistent refrain in the Old Testament, is divine in origin but rooted in an understanding of human nature.

Later on in Genesis, the Bible implicitly records an equation that we have culturally accepted ever since: Wisdom comes with old age. It arrives in the person of Abraham, whom Talmudic scholars often describe as the first biblical exemplar of wisdom being associated with old age. Later still, in the "wisdom literature," we meet Job, that beleaguered archetype of resilience, forbearance, and emotional evenhandedness in the face of divine injustice and relentless adversity. The shorthand lesson of this tale is often conveyed as the "patience of Job," but in the idiom of modern psychology, perhaps a better way of putting it is that Job had an exemplary ability to cope.

Clayton was struck by the fact that many biblical examples of sound judgment resided in the decisions of judges—how they discerned right from wrong, determined guilt or innocence, and then meted out reasonable punishment. Implicit in this exercise of wisdom, just as it was during the golden age of Greek philosophy, was a broad, socially enforced, and morally rigorous sense of communal justice. Judges were typically depicted as the wisest people in ancient settings, and their sagacity rested not only in their ability to discern the truth (often, as in the case of King Solomon, a truth partially obscured by the cloth of human nature), but to render fair yet meaningful judgment. Clayton recognized that the "wise men" in ancient cultures were often judges, and she noticed that the most enlightened examples of justice from the ancient texts relied on a delicate balance of heart and mind, reason and compassion—the rational mind assesses a situation in arriving at a judgment, but the emotional heart attends to feelings when it comes time for justice to be dispensed.

"What emerged from that analysis," Clayton says, "was that wisdom meant a lot of different things. But it was always associated with knowledge, frequently applied to human social situations, involved judgment and reflection, and was almost always embedded in a component of compassion." The Hebrew term for wisdom, *chokhmah,* suggested that the concept resided in both the mind *and* the heart. Indeed, the melding of these two ancient concepts in a single word anticipates one of the burning issues in modern neuroscience: the degree to which mind and body, cognition and emotion, are similarly melded. Not distinct parts, mind you, but marbled together.

At that point, in the mid-1970s, Birren advised Clayton to get her nose out of books and "become more scientific," treating wisdom as a psychological construct that could be defined well enough to be measured and studied. Clayton's fascination with judges and judgment led to a short-lived and, in retrospect, possibly shortsighted exploration of contemporary wisdom conducted by her and some of her colleagues at USC. They wanted to try to measure wisdom in a contemporary group of individuals. And the group they decided to focus on—it seems laughable at first—was lawyers.

There was sound logic to Clayton's approach, however, and she recruited the help of Dorothy Nelson, the esteemed dean of USC's law school, to pursue the idea. In a 1974 letter to participants in the study, Clayton described her thinking:

> Our decision to sample lawyers came after an extensive historical analysis of ancient literature to see in what context the word "wisdom" was used and to which groups of individuals the word was most applied. What consistently emerged was the association of wisdom with a formal educational process that taught men to write and learn the codes and laws of the nation. The men who received such an education were eventually, with age, allowed to sit in the royal court and be a judge. These judges were perceived as wise and were often sought out for advice. Wisdom, then, was an acquired rather than a genetic characteristic. One had to partake of a formal or informal educational process (e.g. strict parental instruction) when young to be wise when old. It seemed that legal education today most closely approximated the process outlined above by emphasizing the skills needed for decision making, by teaching individuals how to ask the appropriate questions to arrive at the most effective solution, as well as by teaching codes and laws.

Even more presciently, Clayton designed her first wisdom study to test subjects at three different life stages of "expertise": law students (whose mean age was about twenty-five), law professors (mean age thirty-eight), and older practicing attorneys (mean age sixty-eight). "Our main concern in the study was to examine the adage that with age comes wisdom," Clayton noted. But as has been true at least since the time of Socrates, there may have been more wisdom in the questions than in the answers: The small study was inconclusive. If anything, the middle-aged group of law professors conformed more closely to our notions of wisdom than the older group of attorneys, who came across as inflexible, with a greater need for deference. The most important thing about Clayton's study, however, is that it brought empirical measures, however provisional and imperfect, to the psychology of wisdom for the first time.

Between 1976 (when she finished her dissertation) and 1982, Clayton published several groundbreaking papers that are now widely acknowledged as the first to have suggested that researchers could bring some semblance of empirical rigor to the study of wisdom. Based on this research, three general areas emerged as central to wisdom: the cognitive, affective, and reflective. Even then, old age appeared to be important to wisdom, but not essential; the more experience you have, she believed, the more chances you have to be wise.

More important, Clayton put the "should" back into wisdom—that is, injected moral ballast into the notion of decision making. As she wrote in 1982, drawing a sharp distinction between intelligence and wisdom, "The function of intelligence is characterized as focusing on questions of *how* to do and accomplish necessary life-supporting tasks; the function of wisdom is characterized as provoking the individual to consider the consequences of his actions both to self and their effects on others. Wisdom, therefore, evokes questions of *should* one pursue a particular course of action." As we'll see shortly, this notion of "should," too, has become a battleground in neuroscience.

As Clayton began to describe her work at psychological meetings in the late 1970s and early 1980s, her wisdom research caused a considerable buzz. Robert J. Sternberg, a prominent American psychologist now at Tufts University, says Clayton's early work was "a big deal. It was a breakthrough to say wisdom is something you could study." Jacqui Smith, a longtime wisdom researcher now at the University of Michigan, called Clayton's early studies "seminal work that triggered all the subsequent studies." One of the people who grasped its significance immediately was Paul Baltes, who then headed the Center for Lifespan Psychology at the Max Planck Institute for Human Development, in Berlin. Baltes closely monitored these initial wisdom studies, Clayton recalled, and regularly peppered her with questions. "I went to all these meetings," she said, "and we would have lunch or dinner at every meeting. He was always asking, where was I with this wisdom stuff?"

The answer would soon be apparent: nowhere. Clayton published her last paper on wisdom in 1982. By then, she had applied for, but failed to receive, a grant from the National Institute on Aging to pursue her study of wisdom, had resigned her position as assistant professor at Columbia University's Teachers College, and left academia for good. Part of the reason, she says, was her distaste for the ruthless nature of academic jousting; and part was recognizing her own limitations in studying such an enormous topic. "I was lost in the Milky Way of wisdom," she admitted, "and each star seemed as bright as the next. Ultimately that's why I didn't continue with it."

The concluding passage of Clayton's last paper focused on a gem of an anecdote that captured the elusive yet central quality of what we think of as wisdom. In an earlier investigation of decision making in an aged population, Clayton wrote, psychological researchers had presented the metaphor of the fork in the road to an elderly woman and asked her if she should take the right or the left branch in the road. After considering the question, the woman finally replied, "I'd just stand at the juncture between the two forks and ask people returning what each path was like."

With Clayton's recusal from the field, the path to wisdom study ran through Berlin, and the working definition of wisdom took on a decidedly Germanic accent.

Like that fork in the road, wisdom lies at the intersection of evaluation and choice—assembling knowledge by whatever strategy (life experience, "book" learning, apprenticeship, interviewing people at the fork), weighing the relative value of all that information, and then arriving at a decision. Psychological researchers were in no position to probe the neural mechanisms of this process; in fact, Clayton explicitly made the point that some aspects of wisdom lay outside traditional cognitive measurement. But psychologists did begin to understand both the kinds of dilemmas that required wisdom and just how complicated, nuanced, and sometimes idiosyncratic such decisions might be.

Here, for example, is an open-ended hypothetical question that researchers used in the Berlin Wisdom Paradigm to assess an individual's wisdom: "A 15-year-old girl wants to get married right away. What should one/she consider and do?"

A wise person, according to the Berlin group, says something like, "Well, on the surface, this seems like an easy problem. On average, marriage for 15-year-old girls is not a good thing. But there are situations where the average case does not fit. Perhaps in this instance, special life circumstances are involved, such as that the girl has a terminal illness. Or the girl has just lost her parents. And also, this girl may live in another culture or historical period. Perhaps she was raised with a value system different from ours. In addition, one has to think about adequate ways of talking with the girl and to consider her emotional side."

An unwise answer? Something like, "No, no way, marrying at age 15 would be utterly wrong. One has to tell the girl that marriage is not possible. . . . No, this is just a crazy idea." The difference between a wise answer and an unwise answer, in other words, lay in understanding context and being flexible.

In the early 1980s, Paul Baltes and a series of collaborators based at the Max Planck Institute for Human Development—a group that eventually included Jacqui Smith, Ursula M. Staudinger, and Ute Kunzmann—embarked on an ambitious program called the Berlin Wisdom Project to, as they put it, "bring research on wisdom into the laboratory." Both Clayton and Sternberg, a psychologist then at Yale University, had created the intellectual space for experimentation in earlier work that had drawn a sharp distinction between wisdom and intelligence, the latter being a trait that had lent itself to measurement (however imperfect) for decades. The Berlin group was "spurred by a motivation to identify and highlight the best of what society and humans can accomplish concerning their own development and that of others." In addition to many papers published over two decades, Baltes wrote and posted on the Internet a book-length work in progress on the subject; entitled "Wisdom as Orchestration of Mind and Virtue," it is one of the most comprehensive (though lamentably still unpublished) treatments of wisdom.

Boiled down to its essence, the Berlin Wisdom Paradigm defined wisdom as "an expert knowledge system concerning the fundamental pragmatics of life." Heavily influenced by life-span psychology, the Berlin version of wisdom required expert knowledge of both fact and human nature; an appreciation of one's historical, cultural, and biological circumstances during the arc of a life span; an understanding of the "relativism" of values and priorities; and an acknowledgment of uncertainty, at the level of both thought and action. "Wisdom-in-action," as they put it, might manifest itself as good judgment, shrewd advice, psychological insight, emotion regulation, and empathetic understanding; it could be found in familial relations, in formal writing, and in the relationship between a student and mentor or a doctor and patient. Although we typically think of wisdom as a personal attribute, the Berlin group saw it as a quality not only of individuals but of groups, institutions, and societies at large; and not just a judgment or action but, also, the *process* that produced the judgment. By its very nature, however, wisdom was a frankly utopian concept and was virtually unattainable. Baltes and Staudinger pointed out in one paper that "wisdom is a collectively anchored product and that individuals by themselves are only 'weak' carriers of wisdom."

And yet individuals must contend with life dilemmas all the time, of course, so one of the most fruitful avenues of research developed by the Baltes group was to measure wisdom by posing open-ended "life hypotheticals," such as the vignette about the fifteen-year-old girl. The Berlin group shrewdly probed for wisdom by using the kind of practical dilemmas many of us recognize only too well: dealing with a relationship in crisis, advising a friend contemplating suicide, making job choices in two-career families, and making life decisions in the face of, for lack of a better term, existential dissatisfaction. In every instance, these dilemmas represent, literally or figuratively, a fork in the road that requires discernment, evaluation, and ultimately choice.

What guides these choices? The Berlin group launched an extensive study of proverbs, in the belief that these little aphorisms captured at least a shorthand folk version of universal truth and wisdom. Baltes wrote a brief history of proverbs, from the earliest Mesopotamian and Egyptian writings around 2500 B.C. to the wisdom literature of Proverbs, in his unpublished book. They ranged from the trope of fathers advising sons about proper behavior ("Marrying several wives is human; getting many children is divine") to biblical advice about living a good (God-fearing) life ("He who spurns his father's discipline is a fool, he who accepts correction is discreet").

At their best, proverbs and aphorisms are like philosophical haikus, quick distillations of universal truths; all too often, however, they are the cocktail peanuts of conventional wisdom, easy to munch on but not very good for long-term sustenance. As Baltes pointed out, many famous maxims flatly contradict one another. But even these contradictions tell us something important about wisdom—as he suggested, it is relativistic and depends on context, so that sometimes the bird in hand is worth more, sometimes two birds in the bush are a better deal, and that rarest of birds, the wise person, somehow knows the difference.

As in any professional subculture, the Baltes group invented a scholarly vocabulary for wisdom studies, and that vocabulary often descended— at least to lay ears—into a semantic quicksand of psychological jargon, rife with implicit theories of wisdom, explicit theories, meta-criteria, operationalizations, and "lifespan contextualisms." They published papers with titles like "Wisdom: A Metaheuristic (Pragmatic) to Orchestrate Mind and Virtue Toward Excellence." Even people in the field were sometimes left scratching their heads. (Perhaps it derives from differences in language or cultural outlook, but the Baltes papers often retreat into a rigid, impersonal quality of language that sometimes seems at odds with its subject matter.)

The Berlin researchers nonetheless obtained the most comprehensive empirical understanding of wisdom by any single group in modern psychology, and Baltes eloquently distilled the meaning and importance of this work in his unpublished 2004 book. In a complex yet supple definition, Baltes wrote that "wisdom deals with difficult problems of the conduct and interpretation of life, includes knowledge about the limits of knowledge and uncertainty, reflects a truly superior level of knowledge and advice, is knowledge that is at the same time deep, broad, and balanced as well as flexibly applied to life situations, requires a perfect synergy between mind and character, is knowledge applied for the well-being of oneself and others, and finally, that, although very difficult to achieve, wisdom is easily recognized when present." With that daunting inventory of qualities, it should come as no surprise that researchers on the Berlin Wisdom Project were among the first to reach what is now a widespread conclusion: There's not a lot of wisdom around. Put simply, we all know it when we see it; we just don't see it very often.

But Baltes recognized the central importance of studying wisdom in the modern era, when more and more people are living to an older age. During midlife in particular, people crave wisdom as they struggle with intergenerational issues (aging parents, rebellious children), as life throws more curveballs at them, as they search for deeper meaning in their lives, and, perhaps most important, as their vantage point on life makes a subtle but momentous shift—instead of thinking of their lives as distance from birth, they begin to think in terms of distance to death. Wisdom, Baltes concluded, "could serve as a prototype for what may be possible in old age and what we can all strive for."

Jacqui Smith, who was collaborating with Baltes on one of his final wisdom papers in the fall of 2006, when he died of cancer at age sixty-seven, told me the study of wisdom continues to struggle with an ongoing and unusual paradox: Although there is immense interest in the topic, its very complexity and definitional fuzziness exiles it to the fringe of academic respectability. "Probably only somebody who has tenure has the luxury of looking at this," she said. But the wisdom research by the Berlin group "moved the field forward by leaps and bounds," according to Sternberg. "Baltes showed that not only could you empirically study wisdom," he said, "but you could do studies that are quite sophisticated."

Some researchers, however, found the Berlin wisdom studies to be abstract, difficult to understand, and cold—too much emphasis on impersonal knowledge, not enough on emotional intelligence. "It's great work, and they've looked at it more closely than anybody else," said Laura Carstensen, a psychologist at Stanford University. "But I think, and one of the critiques other people have had, is that they left emotion out of it. I don't think you can have wisdom without having emotional regulation be a part of it."

Perhaps the more important point is that the early psychological pioneers in the research on wisdom began to lay the groundwork for more rigorous, neuroscientific investigation into the nature of wisdom. If wisdom has a deep emotional (or "affective") component, then current studies on the role of emotion in judgment and the importance of emotion regulation may shed light on how those rare birds among us occasionally achieve wise behavior. And if wisdom represents a kind of expert knowledge, that, too, lends itself to empirical study. As Baltes and Smith pointed out, wisdom as a form of expertise "translated some of the components of wisdom proposed by the ancient Greek philosophers into the language of psychological scientists at the end of the 20th century, who study exceptional performance in complex domains," such as musicians, chessmasters, and medical teams. Everyone agreed that intelligence was not the best predictor of wisdom-related knowledge, but, rather, a combination of life history, knowledge, interpersonal emotional skills like empathy, and what Baltes called "cognitive style."

And what made it difficult to quantify is also what made it so interesting. Unlike intelligence tests, Baltes and Smith once wrote, when you measure wisdom, "there are no correct answers."

Consider another profile in wisdom: "Claire" is in her late sixties, a mother of seven children who lives in Gainesville, Florida. Her life has not been without heartache or tragedy. She grew up poor and her family has never had a lot of money; one of her children refuses to speak with her anymore, and she has been drawn into custody battles and financial imbroglios with in-laws. More significantly, one of her sons was born with cerebral palsy; rather than place the child in a home, as some had urged her to do, she insisted on caring for and raising him at home with the rest of the family. "I would put my healthy kids in a home first," she told researchers in Florida, "instead of putting a baby in there that can't talk for himself." Despite years of challenge (the son eventually died at age thirteen), she was nothing if not positive in her general outlook. "I don't sit around and dwell on bad things," she said. "I don't have the time for it, really. There's so many good things you can do."

Claire is not her real name, but that is how it appears in the psychological literature. Nothing in her remarks recalls the eloquence of Socrates or the worldliness of Montaigne, yet she is arguably one of the few certifiably wise people in the world—"certified" in the sense that she ranked well above average on a "Three-Dimensional Wisdom Scale" developed by Monika Ardelt, a sociologist at the University of Florida in Gainesville.

Ardelt is one of the researchers who felt the Baltes definition of wisdom suffered from a relative dearth of emotional content. In 1990, as a graduate student at the University of North Carolina, she discovered Vivian Clayton's early research, which made emotion a central component of wisdom, and Ardelt began to construct a more elaborate framework for wisdom upon Clayton's foundation. After she joined the faculty at the University of Florida, her mentor reminded her, "We don't have any ways to measure wisdom." As she began to ponder how to do this, it dawned on her just how tricky it is. "You can't just ask people, 'How wise are you on a scale of one to five?' " she told me when we spoke. "Wise people are humble!"

In 1997, Ardelt and her colleagues received a grant from the National Institutes of Health and the National Institute on Aging to develop a psychological test to assess wisdom. She particularly focused on a trait that often appears in many definitions of wisdom: the ability to deal with crises and adversity without getting overwhelmed. As she has noted, "Successfully coping with crises and hardships in life might not only be a hallmark of wise individuals but also one of the pathways to wisdom." Thus, beginning in December of that year, Ardelt and her colleagues began visiting churches and community groups in north-central Florida and asking senior citizens to participate in a "Personality and Aging Well Study."

The participants (180 in all) did not know that one of the purposes of the study was to road-test a series of questions designed to assess general wisdom. Ardelt's working definition leaned heavily on, and expanded upon, Clayton's original concept. Wisdom balanced on three separate but interconnected ways of dealing with the world: cognitive, reflective, and emotional. Hence, a "three-dimensional" wisdom scale, which, according to the custom of psychological tests, was designated by the initials 3D-WS. The cognitive aspect, for example, included the ability to understand human nature, to perceive a situation clearly, and to make decisions despite ambiguity and uncertainty. The reflective part of wisdom dealt with a person's ability to examine an event from multiple perspectives—to step outside oneself and understand another point of view (economists call this quality "other-regarding," while Buddhists often refer to it as "other-centered," but it is a key behavioral trait in several belief systems). And the emotional aspect of wisdom primarily involved an ability to remain positive and minimize negative feelings and emotions. In an initial phase, participants responded to 132 questions that addressed these issues. Later, Ardelt and her colleagues whittled down the list to thirty-nine questions that seemed to capture the elusive concept of wisdom.

Let us concede right from the outset that there is something utterly quixotic about attempting to assess human wisdom on the basis of a self-report test in which subjects agree or disagree with statements like "People are either good or bad" and "I always try to look at all sides of a problem." Nonetheless, the 3D-WS, Ardelt argues, distinguished "how relatively wise older people cope with life crises in comparison to older people relatively low on wisdom." And when the Florida researchers went back and did intensive follow-up interviews with some of the subjects (including "Claire"), a very seasoned, pragmatic, everyday version of wisdom—the wisdom with a small *w,* you might say—emerged in their life stories.

"James," for example, was a seventy-seven-year-old African American. He, too, was no stranger to adversity. After lettering in four sports during high school, he went off to fight in World War II and, after experiencing the horrors of combat, suffered severe depression upon his return to the United States. He acquired an advanced degree and became a successful school administrator, but his marriage fell apart. He was devastated when his mother died. Yet he managed to articulate a grace and calm that seemed to serve him well. "I've had as much bad things to happen as good things, but I've never allowed any outside force to take possession of my being," he told researchers. "That means, whenever I had a problem, I went to something wholesome to solve it." One of the things that helped, he said, was bowling.

Central Casting does not typically dress up the Wise Man in a bowling shirt, but several qualities emerged again and again in older people like "James" who scored high on Ardelt's wisdom scale. They learned from experience—especially negative experiences. They were able to step outside themselves and assess a troubling situation with calm reflection. They recast a crisis as a problem to be addressed, a puzzle to be solved. They took action in situations they could control, and accepted the inability to do so when matters were outside their control. And they were almost embarrassingly positive: When asked about the most unpleasant thing to have happened to them in the recent past, people like "James" were hard-pressed to come up with an answer. As he put it, "I find life beautiful."

Ardelt doesn't take all this too far, and neither should we. "I don't think we have a clear definition yet of what wisdom is," she admits. "I like my definition. The Baltes people like their definition, and Sternberg likes his. There's no agreement on what wisdom is, and that's the fuzzy part. We're not there yet."

One thing that has emerged from these recent studies, however, is that wisdom can arise in mysterious places and unexpected ways. With Ardelt's help, I had an opportunity to speak with some of the people who ranked high on her wisdom scale. "Claire," it turns out, grew up on a tobacco farm in Kentucky, never finished high school, and harbored no greater ambition than to have children. "We're not mountaineers," she told me, "but we *are* hillbillies."

The formative role of adversity brought to mind Vivian Clayton's father, sitting next to his frail mother in London while German bombs rained down around them, celebrating their survival each time with a cup of tea. It made me curious about Clayton, who disappeared from academia in 1982. I managed to track her down through a short item on the Internet, which described a neuropsychologist of the same name who tended bees as a hobby in Northern California. It turned out to be the same Vivian Clayton, and she agreed to meet with me at her office in Orinda on a sunny March morning in 2007, a few hours before seeing her first patient of the day.

Clayton turned out to be a vivacious woman with a soothingly enthusiastic voice. After all the abstraction involved in thinking about wisdom, she had gravitated to more practical ways of assisting older people; she was a geriatric neuropsychologist, helping families and lawyers determine mental competency in older people experiencing cognitive declines (in fact, she helped write the California State Bar manual for making these determinations). She never contributed anything to the wisdom field after 1982, although Paul Baltes continued to send her papers from Berlin for the remainder of his life and Monika Ardelt sometimes consults with her. I asked her if she regretted not continuing in the field, and she said not at all. "I reached a fork in the road," she said. "Wisdom can be a very abstract concept, and as I got older, I gravitated to more practical approaches."

We talked about the absence of wisdom in contemporary culture, especially in politics. "The last remnant we had in our culture of choosing leaders for their wisdom was probably the way Native Americans treated their chieftains," she observed. Then the conversation turned to bees. Perhaps it's a coincidence, but many of the writers we traditionally consult about wisdom, like Aristotle and Marcus Aurelius and Emerson, not to mention sociobiologists, always seem to talk about bees, and it's even spread to the social sciences. Psychologist Jonathan Haidt now speaks of "hive psychology" to describe the positive aspects of human "ultrasocial" cooperation.

"You know, bees have been around for three hundred fifty million years, at least as living creatures," Vivian Clayton said. "And when you work a hive, and you're there with that hive alone, and you hear how *contented* the bees are, you just have the sense that they have the pulse of the universe encoded in their genes. And I really feel that the concept of wisdom is like that, too. Somehow, like the bees, we are programmed to understand when someone has been wise. But what wisdom is, and how one learns to be wise, is still somewhat of a mystery."

PART TWO

EIGHT NEURAL PILLARS
OF WISDOM

Science—good science—becomes a form of wisdom when it's totally disinterested.

—Jean-François Revel

The narrow gate to wisdom lies in science.

—Immanuel Kant

CHAPTER FOUR

EMOTIONAL REGULATION

The Art of Coping

Imperturbable, resolute, tree-like, slow to speak—
such a one is near to Goodness.

—Confucius, *Analects*

ON A BEAUTIFUL SPRING DAY in Palo Alto, California, with rosemary
bushes bristling with purple blooms and lilacs scenting a slight breeze
flowing through the Mission-style colonnade outside Jordan Hall, researchers at Stanford University's psychology department were busy torturing a
gray-haired, good-natured woman named Nancy Lynne Schmitt. Not
that kind of torture. They were, however, subjecting her to a psychomet-
ric version of the Inquisition. For the better part of five hours, Nancy
Lynne sat and squirmed during an endless series of cognitive, behavioral,
and emotional assessments; I felt worn down just sitting in on the
process.

She repeatedly filled out questionnaires asking her to gauge the intensity of more than two dozen emotions. She took a vocabulary test.
She endured a wearying series of tasks designed to assess the quality of
her memory. Before the gauntlet of testing was complete (it lasted two
days), she would also undergo a functional magnetic resonance imaging
(fMRI) of her brain. Every once in a while, she was even asked to gnaw
a piece of cotton until it was saturated with saliva (a test for the stress
hormone cortisol). An upbeat and wryly self-deprecating woman who
wore a red sweatshirt, casual blue pants, and Reebok walking shoes on the
day she donated her brain (temporarily) to science, Nancy Lynne first
inspected and then chewed on the cotton as if it were a canapé.

Since 1994, when she was fifty-four years old, Schmitt has participated in what is known at Stanford as the "beeper experiment." On three separate occasions since the study began more than a decade ago, psychologist Laura L. Carstensen and her Stanford colleagues have outfitted Nancy Schmitt and several hundred other Northern California residents, young and old, with electronic pagers. As the participants go about their daily activities, they receive random beeps during the course of the day, up to five times a day for a week; by agreement, they drop whatever they're doing and fill out a questionnaire describing the emotions they are feeling *at that exact moment,* whether it's happiness, sadness, disgust, or—as you might imagine when they are beeped in the middle of what one female participant described to me as "doing what husbands and wives are supposed to do"—anger.

The questionnaire sessions complement the emotional picture obtained from the "beeper" research, and they can be frustrating, too. For Nancy Lynne, a low point occurred in the middle of a Tuesday afternoon, when she had been asked to perform a maddening mathematical task. Every time Nancy Lynne made a mistake, and she made quite a few, the humorless examiner would say, "Error," and ask her to start over. She became so flustered that she pretzeled her body into a tense ampersand of anxiety and kept repeating, "Gosh, I can't even think. . . ." Later, she confided, "I was almost in tears right after doing those numbers." Helen Lawson, the technician who administered the psychological tests, calmly reassured Schmitt that she had done just fine. And by the time Schmitt completed the final task of the day, which was to rate her emotions on a scale of one (for low) to seven (for high), she had rebounded quite well.

"Happiness is a seven," she said with a triumphant laugh, checking the last box on the questionnaire. "I'm getting out of here!"

There is, as we already know, a fair amount of disagreement about the definition of human wisdom, to say nothing about whether it increases with age or not. But there is a growing consensus that a lot of wisdom begins with the successful regulation of emotion. This "talent," if you will, is embedded in clichés like "calm under fire" and "cool as a cucumber," but for most of us, it seems like a superhuman task, an internalized Olympics between our ears, to maintain our poise when confronted with any of a thousand quotidian perturbations—an insult by a work colleague, the driver who cuts us off, the needy child, the plaintive parent, the lifelong and seemingly irremediable bad habits of our beloved spouses, or just ordinary momentary frustrations, such as Nancy Lynne's, that poison the well of good feeling and knock us for a loop.

A careful titration of passion and detachment in the face of challenge has long been recognized as one of the enduring qualities associated with wise conduct and leadership. It requires balance (Heraclitus: "To be evenminded is the greatest virtue") and self-awareness ("He that can compose himself," said Poor Richard, "is wiser than he that composes books") in the service of emotional resilience. In a broad sense, that is precisely the message that has emerged over the years from this well-controlled, detailed longitudinal study at Stanford, which is unusual because of its power to compare the emotional flexibility of young and old adults, and to chart how they change over time.

What the Stanford researchers have found in their laboratory tests and their brain scans and their random beeper checks out in the real world is that despite the well-documented cognitive declines associated with advancing age, older people in general seem to have figured out how to manage their emotions in a profoundly important way. In general, they experience negative emotions less frequently than younger people, exercise better control over their emotions, and, as Nancy Schmitt did after experiencing frustration during one of the psychological tests, rely on a complex and nuanced emotional thermostat that allows them to rebound quickly from adverse moments. In short, they take things in stride, and quickly return to a state of equilibrium after episodes of upset. In achieving emotional balance, emotional self-awareness, and emotional resilience, they have mastered the art of coping.

But how? What's their secret? Carstensen and her colleagues have proposed that successful emotional regulation is tightly connected to a person's sense of time—usually, but not always, time as it is reflected by one's age and stage of life. "According to our theory, this isn't a quality of aging per se, but of time horizons," she explained. "When your time perspective shortens, as it does when you come closer to the ends of things, you tend to focus on emotionally meaningful goals. When the time horizon is long, you focus on knowledge acquisition." The working hypothesis—Carstensen calls it "socioemotional selectivity theory"—suggests that in the shortened time perspective that comes with aging, people are motivated to focus on what is most important. At a time of life known for cataracts and macular degeneration, they see the truly important things very clearly.

Emotional resilience has long been a compelling trait, in the wisdom literature as well as in the psychological literature. Many researchers are familiar with a classic case study of emotional adversity that involved a middle-aged farmer who suffered a devastating and well-documented series of physical and psychological setbacks. The man lost all his livestock to poachers and rustlers; soon after, a freak tornado struck the home of his eldest son during a family gathering, killing all ten of the farmer's children at once; soon after, he developed a rare dermatologic condition that left his body covered in sores and scabs from head to toe. He lost his sense of taste, his appetite for food, and, ultimately, his appetite for life. As he struggled with bitterness and a familiar "Why me?" sense of victimhood, even his closest friends abandoned him, essentially blaming the victim for his misfortune. The name of this farmer of course is Job, and his travails form one of the richest narratives in human suffering.

In the marketplace of cultural mythology, the story of Job is usually packaged as a tale of patience, but this misreads (indeed, sedates) the full grandeur of Job's emotional lamentations. He himself scoffed at the idea of patience ("Why should I not be impatient?" he cried at one point), and his angry brief against divine injustice is at turns bitter, uncomprehending, mournful, sarcastic, philosophic, and disillusioned. His complaints against God were dismissed as foolishness by his erstwhile companions. "It is not the old that are wise," the young firebrand Elihu told him, "nor the aged that understand what is right."

And yet the single quality that stands out throughout Job's ordeal, the quality that not only that allows him to endure God's injustice but grants him the courage to speak truth to ultimate power, is emotional resilience. He bends, but never breaks; he admits to enormous sadness and bitterness but always returns to a set point of emotional equilibrium and courageous self-regard. "I hold fast my righteousness," he says at one point, "and will not let it go; my heart does not reproach me for any of my days." It is this resilience, not patience, that elevates Job to an exalted perch in the Bible's wisdom literature, and why his story has relevance for our modern notions of wisdom. He summed it all up in five words: "Surely vexation kills the fool."

The role of emotion in wisdom probably seems obvious, but the fact that wisdom "belonged" to philosophy for at least twenty-four centuries meant that a huge edifice of pretty sophisticated argument shored up the notion that rational thought reigns supreme; with a few notable exceptions (especially the eighteenth-century British philosopher David Hume), emotion has traditionally been viewed not as informative, but, rather, as either inconsequential or, more often, a saboteur of pure reason.

But emotion began to assume a more concrete, physiological, deeply biological aspect of human behavior in the latter half of the nineteenth century in the well-known psychological writings of William James and, somewhat less well known, in the natural history of emotion that Charles Darwin assembled. In his groundbreaking 1884 essay "What Is an Emotion?" James suggested that feelings like anger or fear were essentially embedded in the body's "sensorial" apparatus, at a time when those senses were being traced back to the brain. A dozen years earlier, Darwin had tackled the issue of "expression"—that is, the physical manifestation of emotions like disgust, sadness, and anger—in animals and humans, again suggesting that they were not only essential aspects of animal behavior but, in fact, were highly visible facial advertisements for a nervous system in some form of arousal.

That is why, when psychology began its more formal academic assessment of wisdom, Vivian Clayton's work was so influential. From the very beginning, she insisted on the importance of emotion in any definition of wisdom. She also stressed that compassion (the subject of a separate chapter) formed a cornerstone of wise behavior. Coming from the other direction, scientists like Antonio Damasio, Joseph LeDoux, and Jonathan Haidt have popularized the growing scientific consensus that emotion not only colors our perceptions and decisions; it appears, at the level of neurological circuitry, to be deeply embedded in the machinery of thought, wielding its influence at an imperceptible subconscious level and to a much greater degree than anyone had previously imagined. As LeDoux once told me, "Consciousness is the weird thing, the unusual thing."

James J. Gross, whose office is in the same building at Stanford as

Carstensen's, has pointed out that *regulation* of emotion has been an area of keen psychological interest since Freud and his concept of defense mechanisms. But it has been only in the last decade or so that neuroscience has begun to view it as something that can be studied. In a chapter in the latest *Handbook of Emotions,* Gross describes five different strategies of emotion regulation, each with different consequences, and notes that the "crucial" issue is figuring out which strategy best promotes one's goals (is there a better thumbnail sketch of emotional wisdom?). Gross and his colleagues have conducted a fascinating series of experiments in recent years, identifying some of the neural mechanisms by which we regulate emotion. In one recent study, for example, they showed that people who experience negative emotion can diminish the effect by a process Gross calls "reappraisal"—an activity that reveals heightened activity in the cognitive area of the brain (the prefrontal cortex) and dampened arousal in the amygdala and insula, two areas active in the processing of emotion.

"Reappraisal," as Gross sees it, is something that happens very quickly, in a matter of seconds. Yet it requires cognitive intervention, and we might see it, if we expand our aperture, as a process related to reflection, or contemplation. Indeed, psychologists are beginning to appreciate that although the emotional system acts with lightning speed, the cognitive part of the brain can intervene within seconds to blunt affect, and can continue to regulate emotion over a long period of time. Gross points out that "rumination"—the idea of turning over a thought again and again—used to have a purely negative connotation in the psychological literature, and was a risk factor for depression and anxiety. "There's an emerging understanding," he said in an interview, "that there are two forms of rumination. There's the bad form, where you're thinking and thinking and brooding over something, and there's a good or healthy form called reflection. Reflection has more to do with where you are deploying your attention, and reappraisal has more to do with where you are putting your cognitive efforts to affect the past, present, and future." The timescale for these cognitive strategies of emotion control can range, according to Gross, from seconds and minutes to "weeks, months, and even years." Wisdom, he added, "all goes back to being skillful at engaging, in various forms, these strategies of emotion regulation."

What neuroscience has *not,* for the most part, addressed with any quantitative or intellectual vigor is how emotion regulation might change over the course of a lifetime, with its implicit promise that we might become more evenhanded (if not wise) over time. That is where life-span developmental psychology has been so instructive.

Laura Carstensen's professional epiphany on this point grew out of a common assumption that turned out to be wrong. While doing her dissertation at the University of West Virginia (she worked under Paul Baltes and later spent time as a fellow in the Berlin group), she decided to focus on social isolation in the elderly. Like everyone else, she embarked on this research with the conviction that isolation was bad for old people.

But context is everything (especially when it comes to wisdom), and when she conducted studies on the residents of nursing homes, her assessments of these people stood that idea on its head. Coming theoretically from the field of gerontology, which presumed pathology in those who were socially isolated, she kept asking, What's wrong with them? The answer turned out to be nothing. In fact, the people who interacted the least with other social partners in the nursing home setting rated the highest in neurological assessments.

"In one study, I remember, we'd get the nurses to nominate those people who were socially isolated and those people who were more active," Carstensen recalled in a recent conversation in her office at the Stanford Center on Longevity, where she serves as director, "and then follow them around, observe them, write down what they're doing, and do some cognitive testing. And it turns out that the people who interacted the least were the people with the highest cognitive functioning and the ones who were most reluctant to get engaged in the nursing home world."

Intrigued by these findings, she started to ask these people, "Why aren't you out there doing more?" And they would say things like, "I don't want to talk to these people. Have you been out there? Have you seen what's out there?" When Carstensen pressed them further, "they started talking about time and saying things like, 'Well, I don't have time for those people.' Over the years, I moved away from doing my research in nursing homes, like much of the field did, but people still talked a lot about time. And it finally dawned on me that they weren't talking about time in the day. They were talking about time left in life. And then a lot of things began to fall into place.

"This isn't about age," she said after a pause. "This is about time."

The notion that infirmity, loss, the pinch of time, and impending mortality might affect a person's sense of emotional balance may not seem like a particularly earth-shattering insight, yet the importance of the passage of time has been acknowledged much more by the wisdom literature of psychology and philosophy (as well as in the work of Shakespeare, Donne, Keats, and, for that matter, Randy Pausch) than by contemporary neuroscience.

Baltes specifically called attention to a momentous conceptual shift in our worldview when we reach midlife, a kind of temporal continental divide. "As we approach old age," he wrote, "distance from death emerges as a stronger component of our time perspective. As we deal with this change in our conception of 'lifetime,' as we count the years to live more than the years lived, the pressure to set priorities and to re-evaluate the meaning of our lives increases." In a parallel vein from the field of contemporary philosophy, John Kekes identified physiological and biological changes over one's life span as among the limitations that critically shape our notion of wisdom. "Mortality is only one of the commonplaces of whose significance wisdom is a reminder," he writes. "Other limitations are imposed by one's physical capacity, health, temperament, emotional range, talents, society, culture, historical period, and so on. To be wise is to be alive to these limitations in the construction of one's pattern and to be foolish is to be blind to them."

"I'm not a wisdom person," Carstensen felt obliged to tell me. But she readily agreed that many elements of emotional regulation seen in older adults are "absolutely consistent" with qualities that have long been identified by psychologists and philosophers as core components of wise behavior. "We think what happens is that when people are reminded of the fragility of life, people are better able to see what's important and what's not," she continued. "And they see that very starkly, very clearly. Because it's *right now*, what matters now. For most people, those are emotional [goals], emotionally meaningful goals."

Perhaps the most surprising scientific destination for this train of thought is that the biological underpinnings of emotion regulation—

how we cope, how we maintain emotional equilibrium—might possibly be glimpsed in brain experiments. Over the past fifteen years, Carstensen has produced a substantial body of research, blending traditional psychological and behavioral data with neurophysiological measures and brain scans. Her group has shown that the ability to focus on emotional control is tightly linked to a person's sense of time. Young people, who tend to have an open-ended sense of the future, typically pursue what Carstensen calls a "knowledge trajectory"; they are interested in careers and the acquisition of information (and, brain studies show, they seem to hold on to negative emotion more tightly). Older people, as a rule, have a shorter time horizon. Confronted with their own mortality and that of family and close friends, they more typically pursue a trajectory that emphasizes emotional richness and social connection. And these differences correlate with important features of brain activity and overall health.

The results of the "beeper experiment," for example, confirm that older people are, on average, more even-keeled emotionally (Carstensen sometimes uses the word *poignant* in an attempt to capture the bittersweet balance of optimism and loss that colors their lives). In some ways, Carstensen's research turns Shakespeare on his head. All those sonnets devoted to the idea of carpe diem, of living for the moment, as the birthright of headstrong youth? A wealth of research, including the most recent results from the beeper study, argues that with shrinking time horizons, older people are the real masters of carpe diem. They have learned how to extract the maximum emotional satisfaction from life while minimizing the amount of energy squandered on negative emotion.

"Younger people are much more negative in their daily life than older people," Carstensen says, "but older people are more likely to experience mixed emotions, a combination of pride and sadness, at the same time. Having mixed emotions helps to regulate emotional states better than extremes of emotion. There's a lot of loss associated with aging, and to the extent that you use these losses as a motivation to savor the day-to-day experiences you have, it allows you to be more positive. Appreciating the fragility of life helps you savor it."

It's no accident that Carstensen's insights took root in the nurturing intellectual soil of life-span psychology, because that's about time, too. When she was a graduate student, the psychology department at West

Virginia included people like Baltes, Warner Schaie (who collaborated on some of Vivian Clayton's early wisdom research), and John Nessel-roade, all of whom are founding figures in the field, and it gave Carstensen a framework to examine how time horizons—long or short, depending on age, obviously, but also on less obvious variables like life circumstances, health, and even traumatic public events of historic dimension such as the Great Depression or World War II—change the lens of one's emotional worldview like a new prescription from a spiri-tual optometrist. Having found that emotional evenhandedness accom-panies a shrinking time horizon, Carstensen is quick to add that she doesn't regard her work as part of a recent research trend known as posi-tive psychology, which associates positive attitudes and virtues with emotional well-being. Baltes, she points out, always insisted that devel-opment entails a shifting emotional ledger of gains and losses.

"When the future is vast and open-ended, people *need* to adapt to that temporal context by preparing for that future," Carstensen says, "and that means you collect—you collect people, you collect experi-ences, you collect information, and you bank it, because at one point in time it may become relevant, even if it's not relevant today. But it's very much this preparatory state that people are in. And in that state, infor-mation is *so* valuable, and any information is valuable. You have to be able to digest it, to learn it, to remember it, to encode it, to build on it, even if that has costs to your emotional well-being."

It *does* have costs. As an example, she mentioned one of the paradig-matic events that test the emotional mettle of young people—whether or not to attend a party. Young people, she said, often feel compelled to attend social gatherings, in part to meet new and interesting people, expand their circle of social contacts, perhaps encounter a potential mate. But attending a party also entails considerable emotional risk in terms of potential embarrassment and blows to self-esteem (you might, for example, see someone in whom you had a romantic interest nuzzling up to somebody else).

"There's lots of negative emotions associated with that sort of entry into the world," Carstensen says. "But you've gotta do it. Because you're preparing, and that's your world and you need to know it. As people per-ceive that future as more limited, they don't have to do those things any-more. The future is now here. What you need for the future is what you have. You don't need to pick up information or people that might be irrelevant. You know what's relevant; you know what's not. And people focus on the present, the here and now. When you do that, one of the most important functions is how you feel in the here and now. Things become much clearer, because people are letting their feelings navigate what they do, whom they spend time with, what are the choices they're making in life. It's about *right now.* So I think knowledge-related goals are chronically activated when people are young, and there's this shift over time to emotion-related goals that become chronically activated. They can become overridden, in both cases. Younger people at times focus on the present; older people focus on the future. But, given no strong opposing forces, people will tend to go in those directions."

Carstensen and her colleagues have actually seen glimpses of this differential, age-related emotional processing in recent studies. In one brain-imaging study of how the brain anticipates loss, the group showed that heightened arousal of the anterior insula in anticipation of monetary loss predicted an individual's ability to learn how to avoid losses in the future. In a separate study examining attitudes toward winning and losing money, Carstensen, Lisbeth Nielsen, and Brian Knutson showed that younger adults experience wide swings in emotion when anticipating either gains or losses while playing a monetary game; older adults expended most of their emotion in anticipation of winning, with little emotional regard for loss. The researchers suggested that older adults may minimize their anticipation of loss "precisely to avoid emotional swings," and go on to add, "Indeed, it is tempting to speculate that accurate affective forecasting"—that is, not exaggerating or suppressing emotional predictions about the future—"may constitute a core feature of wisdom." The results complement earlier brain-scanning experiments done as part of the beeper experiment, which showed that young people flash neurological "concern" in the amygdala at the prospect of monetary loss, while older people show little or no activation in anticipation of loss. It is as if young people become flustered by the possibility of a negative outcome in the future, whereas older people have learned not to squander their emotional capital on everything that *might* go wrong.

"Very consistent with positivity," says Carstensen, "very consistent with regulation [of emotion]." This live-for-the-moment, "carpe diem" effect seems to occur whenever a person's time horizon shifts, and is not strictly tied to one's age; it has already been detected, for example, in young people exposed to life-altering public events, like the September 11 attacks in the United States and the outbreak of the SARS epidemic in Asia.

These findings reinforce a widely accepted (although largely anecdotal) conviction in the field of neuroeconomics that young and old people generally approach decisions (at least economic ones) in different ways. Young people tend to be steep temporal discounters (that is, they want rewards like cash immediately, and don't want to wait for potentially greater rewards in the future), have more of an appetite for risk, and don't much care for ambiguity; by contrast, older people are thought to be shallow temporal discounters, more averse to risk, yet less averse to ambiguity. Neuroeconomists like Paul Glimcher of New York University agree that emotional regulation is a key component in this age-related difference in outlook. "There's no doubt," he said, "that when people are emotionally aroused, they are superdiscounters. And when they are emotionally aroused, they are more loss-averse. They have all these interesting features. And you can actually teach people how to meditate so they don't do that." Indeed, a recent experiment by Elizabeth A. Phelps, a leading neuroscientist at NYU, demonstrated that adults who practiced cognitive strategies of emotional regulation were able to overcome their natural aversion to loss and perform better on decision-making tasks involving economic choices. They had learned, in the words of the researchers, to "think like a trader."

The effect of aging on emotional maturation—and, by extension, emotion regulation—has been relatively neglected by recent neuroscience, which focuses almost exclusively on college-age experimental subjects, but that appears likely to change. The National Institute on Aging (NIA) recently began to fund neuroeconomic studies that compare the responses of old and young people. As Lisbeth Nielsen, who oversees the NIA program (and recently collaborated on a study with Carstensen), dryly pointed out, "The really important decisions we make economically are not taken when we're in college."

But do these people who regulate their emotions, who have this even-keeled ability to focus on the positive and minimize the effects of the negative, tell us anything about wisdom? "People seem to *know* that they are not going to get worked up about negative future outcomes," Nielsen said, "which suggests that they know *how* their life is changing. What that has to do with wisdom is still uncertain. Wisdom could simply be managing your life in this satisfying, nonupsetting way, where well-being comes from maintaining good social relations."

Like ballast in a storm-tossed boat, emotional regulation can't prevent the dips and swells that occur during moments of crisis in our lives, but it may well speed the return to emotional equilibrium, and for that reason, it seems to help people navigate the day-to-day turmoil of life. This sense of boundary and balance, of not getting too low or going too far, has been noted by everyone from Montaigne ("I do indeed lose my temper in haste and violence, but I do not lose my bearings to the point of hurling about all sorts of insulting words at random and without choice") to the anonymous author of Proverbs 12:18 ("Rash words are like sword thrusts, but the tongue of the wise brings healing"). But these philosophical and theological intuitions now find concordance with scientific findings. Fredda Blanchard-Fields of the Georgia Institute of Technology, for example, whose work will be described in greater detail in a chapter about wisdom and aging, has turned out an impressive series of studies showing that the emotional equilibrium of older people allows them to be much nimbler problem solvers than younger people when it comes to interpersonal dilemmas. "She wouldn't call it research on wisdom," Carstensen told me, "but I would."

Richard J. Davidson, a neuroscientist at the University of Wisconsin, has long been interested in assessing the way people regulate their emotions, and he has taken that search into the brain, using both functional MRI scanning and sophisticated, customized electroencephalographs—Medusa-like headdresses of coiled wire and electrodes attached to the scalp—to measure minute variations in brain activity. In a recent study, for example, Davidson and his colleagues looked at patterns of brain activity associated with optimal emotional regulation in a small group of older people who have participated in the Wisconsin Longitudinal Study (this long-running study has attempted to follow, demographically and psychologically, a random sample of more than ten thousand men and women, all of whom graduated from Wisconsin high schools in the year 1957).

In a paper published in 2006, the Wisconsin team reported that older adults (the average age was sixty-four) who regulated their emotions well showed a distinctly different pattern of brain activity from those who didn't. Indeed, the pattern seemed to reveal a conversation going on between different parts of the brain, which, when weighted in one direction, kept negative emotions like anxiety, fear, and disgust in check. These even-keeled people—Davidson specifically refers to them as "emotionally resilient"—apparently used their prefrontal cortex, the front part of the brain, which governs reasoning and executive control, to damp down activity in the amygdala, those twin almond-shaped regions deep in the brain that process emotional content. In people who are unable to regulate their emotions, amygdala activity is higher and daily secretion of the stress hormone cortisol betrays a pattern associated with poor health. "Those people who are good at regulating negative emotion, inferred by their ability to voluntarily use cognitive strategies to reappraise a stimulus, lead to reductions in activation in the amygdala," said Davidson. He added that such regulation probably results from "something that has been at least implicitly trained over the years." In other words, these people have somehow *learned* to regulate their emotions.

Remember this particular axis of neurological activity, this running conversation between the prefrontal cortex and the amygdala, because its significance has emerged numerous times in recent studies of emotional regulation. Several years ago, Carstensen and her colleagues, including neuroscientist John Gabrielli (now of the Massachusetts Institute of Technology), did fMRI studies of young and old people to see if their ability (or inability) to regulate emotions left a trace in the amygdala. Brain scans indicated that when people look at negative images that show, for example, scenes of violence or suffering, both young and old people respond as expected, with a spike of arousal in the amygdala. But young people tend to cling longer, neurologically speaking, to bad news; the amygdala remains aroused longer. Older people seem better able to shrug it off, move on, and shift their focus more to positive images. This selective focus on the positive is, in a sense, a choice guided by emotion, and it again echoes some very old thoughts on wisdom. In modern neuroscience parlance, wisdom may in part be a function of cognitive attention. The ability to maintain emotional balance, and to ignore extraneous or emotionally disturbing information, appears to be strongly correlated with the focus that often accompanies contemplation or reflection.

James Gross, at Stanford, has even pushed this area of research into the realm of gender in an effort to explore whether men and women regulate their emotions differently. His group published a report in 2008 suggesting that men expend less cognitive effort to manage negative emotion (possibly because they tend to regulate emotion automatically), while women use positive emotion to a greater extent to reappraise—again, cognitively process and modulate—negative feelings. The study, however, was small (twenty-five subjects) and limited to people between the ages of eighteen and twenty-two, so it's a little too early to say that the "women from Venus, men from Mars" cultural metaphor has gone molecular.

There have been other intriguing experimental findings that support some sort of "positivity" effect, and one recent neuroscience study, typically small but quite surprising, has even suggested a role for emotion in human imagination. The experiment investigated how emotion colors the way we imagine the future—which, as we'll see in a later chapter, is a crucial component in both willpower and wisdom.

Over the past three decades, psychologists have documented what they have called an "optimism bias." In tests of a broad range of activities, studies have repeatedly shown that people expect to live longer, be healthier, have greater success in the job market, and avoid the likelihood of divorce than the actual data would suggest. Indeed, many of these expectations about the future are plainly inaccurate, according to researchers, and yet this bias toward optimism—this uplifting and energizing *delusion,* if you will—is pervasive. Knowing that emotion is bound up in the memory of past events, Tali Sharot, Elizabeth Phelps, and their colleagues at New York University turned the arrow of time in the other direction. They wondered if emotion played a role in the *imagination* of future events.

Phelps says that the optimism experiment, which ended up being published in *Nature* in November 2007, began as a pilot study in which Sharot asked volunteers to imagine dozens of imaginary future events, such as visiting a museum or ending a relationship, and then to attach an emotional value (positive, negative, or neutral) to this imaginary future

event. To Sharot's surprise, almost all these scenarios tended to elicit positive feelings—even the ones that seemed nominally negative, like ending a relationship. Some participants viewed breaking up as a positive
event, for example, because it created the possibility of a more satisfying
subsequent relationship. On the basis of these unexpected findings, the
NYU researchers organized a more formal study, in which a small group
of young volunteers (fifteen college-age individuals, average age about
twenty-three) had their brains scanned while contemplating events that
had taken place in the past or might in the future.

What they found was that a "moderate optimistic illusion" about
future events—not blindly optimistic, and not unduly pessimistic, but a
"just so" optimism—could serve as a healthy motivation (in terms of
both physical and mental health) toward the attainment of future goals.
In fact, they succeeded in detecting a residue of this optimism in the
brains of their experimental subjects with fMRI scans. The circuitry of
this forward-looking optimism involved cross-talk in the brain between
the amygdala, that powerful locus of emotion, and a part known as the
anterior cingulate cortex, which is located deep along an inner fold of
the frontal cortex. The activity in these two neighborhoods of the brain
was highest in the people who reported the most vivid imagination of
future events. The findings suggest that the brain may be built with an
ever so slight tilt: a bias toward optimism. This may make both behavioral and evolutionary sense, because a belief that a good outcome may
result in the future from present-day action is a powerful motivation for
people to take action in the first place.

It is tempting—and I won't decline the temptation—to see traditional elements of wisdom embedded in this brain circuitry. If, as the
Phelps study suggests, the same neural mechanism that binds emotion to
the recollection of past events is deployed to imagine future events, we
can see how a lifetime memory bank of good or bad emotional experiences can color our view (rose-tinted, or through a glass darkly?) of the
future. And if our thoughts about the future are positive because they
are, as Phelps and her colleagues delicately phrased it, "not constrained
by reality," we can see the wisdom in being mildly optimistic about the
future. *Mildly*, however, is a crucial qualification; as the NYU researchers
noted, extreme optimism can be dangerous, because it can obscure the
clear-eyed perception of risk and encourage poor planning, just as

extreme pessimism can exaggerate risk and constrain possibility. Perhaps mild pessimism, by the same token, also provides the clear-eyed, skeptical view of human nature that has always been a hallmark of wisdom.

The overall point is that the "moderate optimistic illusion" may have some evolutionary value to our species. As Phelps and her colleagues wrote in *Nature,* "Expecting positive events, and generating compelling mental images of such events, may serve an adaptive function by motivating behavior in the present towards a future goal." Put another way, positive imagination gets us out of bed in the morning and pushes us toward the realization of our dreams. Maybe that's what motivated Job to arise each day and, despite all his tribulations, joust some more with God.

When we think about the natural history of wisdom, we eventually have to ask the "E" question. Is there any evolutionary value to the apparent emotional evenhandedness that more mature adults in particular are able to achieve? Does emotional support and caring improve the survival of one's genes? Laura Carstensen and Corinna Lockenhoff recently tackled precisely those questions, especially in light of the well-documented cognitive challenges of old age.

"Information processing capacity declines over adulthood," Carstensen admitted, "but knowledge goes up. So people are learning more. They're not quite as efficient at learning, but the longer you live, you're still learning more. Even into their seventies and eighties, you're seeing people's vocabulary increase, their knowledge about the world increase. In an evolutionary context, there wasn't a whole lot new you had to learn once you reached adulthood, so maybe there weren't selection forces for *processing* information. But knowing how to *do* things would have had much survival value. So to the extent that older people in your group are knowledgeable, evenhanded, and selective—selective theory being that people focus on the most important people or things in their life—those three things together would really increase the survival of that person's genes in a group." Put simply, emotion regulation helps us focus on what's truly important, which is essential not only to wisdom but also to survival.

As Darwin famously noted in *On the Origin of Species,* "Individuals having any advantage, however slight, over others would have the best chance of surviving and procreating over other kinds." So where does that leave us baby boomers and fellow-traveling elders who are past reproductive age? Carstensen and Lockenhoff point out that natural selection doesn't give a hoot about the acquisition of new information late in life. Older adults, they argue, have evolutionary value only if they contribute to the reproductive success of younger *kin*.

But that is exactly what grandparents do, in myriad ways. According to the well-known "grandparent hypothesis," older relatives do this by transmitting their world knowledge and using their higher-level cognitive skills to enhance the lives of their grandchildren. In other words, they *care* for their blood relations, and this emotionally honed attentiveness to the next generation—psychologists like Erikson called it "generativity"—seems to confer survival value. And this applies not just to humans, by the way. Two studies of monkeys have shown that the presence of a "grandmother" increases the reproductive success of the younger kin.

As Carstensen and Lockenhoff write, in a passage that seems as rich in wisdom as in biological observation, "evolutionary selection should have *favored* skills that help older people help others. From this perspective, findings that old age is characterized both by large stores of knowledge about the world and everyday life, and social and emotional investment in younger kin, are hardly paradoxical. Postreproductive adults would be expected to increase their descendants' chances of survival if they were emotionally balanced, knowledgeable about social relationships and the world in general, and invested in social cohesion."

To echo an earlier remark: Carstensen might not call this wisdom, but we surely can. The lifelong accretion of expert knowledge, emotional control, social caring, and insight into human nature arguably increases our odds of survival—as blood relatives (through grandparenting), but also as a species. Emotion regulation may be the most powerful lens in human psychology; polished by time and curved by intimations of mortality, it allows us to see what is really important in our lives.

CHAPTER FIVE

KNOWING WHAT'S IMPORTANT

*The Neural Mechanism of Establishing Value
and Making a Judgment*

> *Judgment is a tool to use on all subjects, and
> comes in everywhere. . . . It plays its part by
> choosing the way that seems best to it, and of a
> thousand paths it says that this one or that was
> the most wisely chosen.*
>
> —Montaigne, "Of Democritus and Heraclitus"

PAUL GLIMCHER IS AN INTELLIGENT, gregarious, perfectly pleasant
and seemingly healthy man with a textbook case of multiple personality.
He insists that he is three different people, each with a unique and sometimes contradictory view of the human mind and how it makes decisions.
One of the three Pauls is an economist. The second is a psychologist.
And the third is a neuroscientist, trained at Princeton and the University
of Pennsylvania, who heads the Center for Neuroeconomics at New York
University. His symptoms include auditory hallucinations: He admits to
hearing voices (usually the voice of Nobel Prize–winning economist Mil-
ton Friedman), which whisper sweet libertarian reminders in his ear
about human preferences and human decisions.

Glimcher claims no expertise in wisdom, but he is one of the leading
experimentalists in the emergent field of neuroeconomics, and his laboratory at NYU, with one branch dedicated to studying the neurons
involved in decision making in animals and another welcoming a steady
stream of human guinea pigs to an fMRI suite near Washington Square,
is at the forefront of deciphering what goes on in our brains when we

make a decision. The choices he studies are not necessarily good or *wise* decisions, I should add, but they are part of a global research effort to identify the mechanics of judgment, and as such, they are giving us a first glimpse at the basic machinery of wise decisions.

My observation about his "multiple personalities" is facetious, of course, but also instructive, because when Glimcher speaks about how humans make choices (usually, in his experiments, rapid choices about modest economic rewards), he freely admits his own internal tensions while thinking about these issues as an economist, a psychologist, and a neuroscientist. More to the point, his tension is ultimately our tension, and it can help clarify for nonscientists what these experiments can, and cannot, tell us about discernment, evaluation, reflection, and judgment—the invisible process of weighing the relative value of different courses of action that goes on inside our heads, often unconsciously and with blazing speed, before we make a decision.

Glimcher's lab has recently explored a genre of experiment that has by now become a cliché in science (and, alas, science journalism)—offering people a choice between, say, twenty dollars immediately or forty dollars in two months, and then measuring activity in the brain as people grapple with this small economic conflict. In a larger sense, his research is teasing out the specific neural circuitry involved in the way we weigh information, the way we evaluate possible choices, and how we deliberate before making a decision. And lurking behind that is a meta-issue: What makes a judgment shrewd, good, even wise? And do our choices even reflect "choice" at all, or are they the inevitable conclusion of neural processes that amount to very precise, physiologically based algorithms about what is "important" lodged in neurons?

A lot is up for grabs in this research—not just the neural underpinnings of something as grand as wise judgment but also issues of free will, conscious choice, and even the possibility of training our minds to be more efficient about optimizing our decision-making ability, essentially by gaming the neural circuitry that is beginning—repeat, *beginning*—to come into focus. So these almost shamelessly simple economic experiments are freighted with considerable human significance. And some of the science threatens to take us to a humanistically sterile place—sterile not because the scientists are so clinical and uncaring about the inner mysteries of a mind at work, but because those workings challenge some of our most cherished ideas about contemplation, reflection, and judgment.

So I want to embark on this potentially disillusioning journey with one important caveat and one frivolous but thought-provoking analogy to popular culture. The caveat is that, although great progress has been made in recent years in understanding the neural apparatus for decision making, in this and many other laboratories in the United States, Europe, and Japan, the complementary machinery that allows us to attach value to choices *before* we reach a decision—in effect, allows us to establish in our own minds what is most important before pulling the lever in our little neural voting booth—is still very unclear and almost certainly will yield further surprises. Despite everything you've read in books and newspapers and on blogs, the neural circuitry for judgment and decision making is still a work in progress. But it is part of a maturing science that, as philosopher Patricia Smith Churchland rightly predicted more than two decades ago in her book *Neurophilosophy,* will be historically transformative: "In its power to overturn the 'eternal verities' of folk knowledge, this revolution will be at least the equal of the Copernican and Darwinian revolutions."

Now, from Copernicus and Darwin to Hanna and Barbera. "I grew up watching *The Flintstones,*" Glimcher told me when I visited his office recently. "And when Fred and Barney's car would break, they would open the hood and they would throw pieces out. You remember this? And then they would close the hood and the car would drive off without those pieces just fine." He paused just long enough to laugh nervously, and then said, "I find myself sort of in this position. As we're building a model that predicts human behavior at the neurobiologic level, that piece—free will—just does not seem required. I mean, there seems to be no compelling evidence that we need that piece."

Leaving free will, wisdom, and other modest issues aside for the moment, let's look at what's left under the hood in human decision making. What's the basic neural circuitry for making a choice, wise or otherwise? How do we evaluate two competing options to make the best decision? And do these admittedly reductive and artificial laboratory experiments tell us anything about the more profound and important choices that we make in what we might smugly call "real life," like choosing a college or deciding on a career?

. . .

In the fall of 2006 and the spring of 2007, a dozen young college students—the lab rats of contemporary neuroscience—visited and revisited the functional MRI suite in the basement of a building in Greenwich Village that houses NYU's Department of Psychology to participate in what has already become a classic form of neural experimentation. As their brains were being scanned, the students assessed a series of economic offers and then made a choice: They could receive an immediate reward of $20 (and the money was real, immediately deposited in their bank account) or a larger monetary award at some point in the future. In some of the trials, they might have been offered $20.25 if they waited a mere six hours, or $110 if they waited 180 days. (What makes this kind of MRI "functional" is that it measures changes in real time of oxygen-rich blood in various regions of the brain, with the assumption that the parts of your brain that are working hard on a particular problem or task are metabolically more active and thus require more oxygen; for more about MRI, including some caveats, please see the note on page 285.) As the experimental subjects weighed the values of these competing offers, the MRI machine essentially took pictures of their brains as they assessed the value of a bird in hand versus two, or even five, in the bush.

At one level, you could say that the NYU researchers are using a $2 million piece of machinery to give us a breathtakingly detailed neural snapshot of stupidity. Most of the participants in these experiments opt for the immediate reward, even though, from a purely economic standpoint, the wise decision is almost always to take the delayed but larger reward.

"It's bizarre," Glimcher said, shaking his head. "Kid comes into the lab and we offer him a choice between forty dollars in two months and twenty dollars now, and he takes the twenty dollars now. So what have I just offered him? I've offered him a hundred percent interest rate in two months. That's a *six hundred percent* annualized interest rate, and he turned it down! Now, if I put on my economist's hat, it's like, haven't these people ever heard of banks? You know, all you have to do is borrow the twenty [dollars] from yourself, and pay yourself back with one hundred percent interest in two months. You should *always* take the delayed amount. But the fact of the matter is, people do not. It's a robust finding.

Many, many labs have shown this in different settings. And if you explain it to people, they learn to get it right. And, of course, financial institutions get it right." (Or so it seemed when Glimcher made these remarks; several months later, it became clear that many banks had gotten it horribly wrong.)

The essence of the experiment, which Glimcher and his colleague Joseph W. Kable published in the journal *Nature Neuroscience* in December 2007, is that each individual in the study seemed to have a unique (the experimenters actually called it "idiosyncratic") way of assigning subjective value to immediate versus future rewards; a person's sense of time is an essential component in this calculus, which touches upon such basic human behavioral traits as impulsivity and prudence (I'll expand on the effect of time, and what is known as "temporal discounting," in the chapter on patience). And while one could argue that documenting the neural workings of foolishness might shed modest indirect light on wisdom, I want to focus here more narrowly on the nuts-and-bolts mechanics of neural choice. When our brains are evaluating choices—*discerning*, if you will, the relative merits of competing options—what exactly is going on?

In at least a minimalist sense, researchers now have a pretty good idea. "I think that there is no doubt now—in fact, I think all of us agree—that decision making in humans and other primates is mediated by a two-stage mechanism," Glimcher explained. "The first stage is broadly construed as a valuation stage. This is where we set the values on the options that stand before us. The second stage takes that as an input, essentially, and chooses the best option amongst the current choices." The results of the 2007 experiment, and all the others like it, have made clear that this preliminary valuation stage is subjective, idiosyncratic, and, from the point of view of wise economic choice, often deeply flawed. And yet this process appears to be a proxy, a neural sketch, of what our brains do all the time when they evaluate, deliberate, and ponder *any* kind of problem, big or small—processes that inform the exercise of judgment. So how does it work?

The road to understanding how the brain weighs options began back in the late 1980s and early 1990s, when neurobiologists uncovered a neat cellular mechanism that seemed to undergird both learning and a sense of reward. It is called "reinforcement learning," and the juice that drives it is an intensely gratifying, immensely astute molecule called dopamine. This is often described as a pleasure molecule, but it's really more like a movie critic, assessing the success of our behavior when we seek a reward and broadcasting its opinion thoughout the brain. This neurotransmitter is released, and gives us a quick spritz of assessment, in the very core of our mid-brain when we feel rewarded, whether by food, drink, money, sex, heroin, a pat on the head, or a kind word from strangers. The very phrase "reinforcement learning" unites vastly disparate parts of our biology—"reinforcement" occurs in the emotional core of the brain, the reward center, while "learning" resides in the cognitive, information-gathering part—and, if you will indulge a bit of skywriting here, this system is so enmeshed in the neural mechanics of habit, motivation, procrastination, discernment, patience, adaptation to change, feelings of self-satisfaction, and, most of all, the attachment of value to things (which is a nonphilosopher's way of deciding what is important) that, I would argue, it's impossible to think about wisdom in the twenty-first century without at least acknowledging the fundamentals of this network of neural plumbing. Not the whole story, I should add, but the neurophysiological *framework* for the story.

Here's an example of how it works, courtesy of a computer game Glimcher's student Robb Rutledge developed and uses in experiments. It involves fishing for crabs, and I had an opportunity to play. You look at a computer screen and see two buoys—one on the right, which is red, and one on the left, which is green. When you click on one or the other buoy, a crab cage rises out of the water, either empty or filled. Your job—one any fisherman would understand—is to figure out where the crabs are and harvest as many as possible.

At first you proceed by trial and error, moving between one of two buoys to figure out the best place to catch the crabs; when you discover that many of them are congregating around one particular buoy, you become very pleased with yourself. That pleasure has a neurobiological basis; when you achieve a goal, you essentially give yourself a pat—not on your head, but *inside* your head. That reward takes the form of a little spike in dopamine, which gives your decision a good review. Your brain has rewarded your persistence in learning where to find crabs by giving you a quick hit of dopamine in the mid-brain.

So far, so good, but this system is actually more complicated (and, to my mind, more interesting) than onetime gratification. As you return time and time again to the preferred buoy and keep pulling up crabs, you don't get quite the same dopamine jolt as you did initially. You're still pleased with your fishing prowess, and guided by dopamine in your choices, but there's nothing surprising about your success. Your sense of reward has subtly become displaced—it's not finding crabs that is driving your behavior and choices, but the *anticipation* of finding them. In other words, the timing of when the dopamine neurons fire shifts from pure reward to *prediction* of reward. And if you're always successful, the anticipatory jolt of dopamine trickles down to nothing. When outcomes become so predictable and expected that there is no longer any surprise, the dopamine faucet essentially turns itself off. We become neurologically bored.

But we don't live in a static, predictable world, and our brains are built to learn from the unexpected. Back on the bay, let's say that you return to your favorite buoy, expecting success, and fail to catch a single crab (Glimcher's researchers have programmed various patterns of evasion for their animated crabs in order to tease out the brain's response to this fishy form of disappointment). How does your brain respond to this development? Surprisingly—and this indeed created a huge surprise when German scientist Wolfram Schultz first reported it in 1996—the neurons that release dopamine freak out when their predictions are wrong. The system is actually built to detect an error, an unexpected outcome. That's when the brain *really* takes notice.

In an odd way, your brain's appetite for learning is especially stimulated by failure, because an error in prediction is what motivates you to find a new solution to the problem. Several computationally savvy neuroscientists then at the Salk Institute, in California—Peter Dayan, Read Montague, and Terry Sejnowski—realized that the pattern in which these "disappointed" dopamine neurons fired in the mid-brain matched computer algorithms designed to model decision making and choice. So when you continue to search for the right fishing hole, and keep up with those elusive crabs, your strategy over time is shaped by timely squirts of dopamine. In this sense, reinforcement learning is just a molecular way of saying *experience*. A quick way to boil down the implications of this neural plumbing: Success breeds habit and failure breeds learning. We are designed to learn from mistakes, errors, unexpected outcomes. The brain feeds its quest for more and better knowledge by turning on an internal sprinkler and dousing itself in a neurochemical cocktail of motivation when it needs to learn something new.

"Now that's cool," Glimcher said, "because reinforcement learning devices are devices for learning, by trial and error, the values of things and representing them. So this immediately raised the possibility that targets of the dopamine system were at least one of many systems by which trial and error would learn the values of actions."

The neurons that secrete dopamine are especially dense in an evolutionarily "older," inner part of the brain—the basal ganglia, the ventral striatum, and the medial prefrontal cortex (in terms of neural geography, the basal ganglia is a knot of neural tissue toward the middle of the brain; the ventral striatum is at about the level of the lowest part of the earlobes and almost in the center of the brain; and the medial prefrontal cortex is almost like a pair of hands cupped around the striatum). There are several fascinating aspects of the dopamine system. One is that the brain is especially keen about noticing the unexpected; that's what really jangles the dopamine system. And seizing on the unexpected as a motivation creates, again, an immensely powerful engine for learning.

So through learning, the dopamine system—until further notice—essentially encodes value and passes those predictions of values to circuits that make decisions. Those decisions, in the neural sense, are virtually global in possibility—to scratch your arm or to scratch behind the ears of your dog; to turn back and look at a sunset again or to turn back and look at the destruction of Gomorrah; to say everything from "I love you" to "Pass the salt."

Nor is decision making an essentially human trait. Bees make decisions to guide their choices while foraging for nectar (which, after all, is just a variation on fishing for crabs), and there was even a recent article in the journal *Science* on "decision making" in fruit flies when they select the optimal site for laying eggs. What makes it a little creepy in humans—and this gets back to free will—is that Glimcher claims that by using the model of dopamine-driven decision making, he has learned to predict which crab pot a player is going to pick in the "fishing for crabs" game with eyebrow-raising accuracy.

"If I stop you," Glimcher said, "and ask you, 'What are you going to do next?' you'd say, 'No one could predict it. I am asserting free will.' But the fact of the matter is that our model would predict ninety percent of your choices at this point. And from a sample of a hundred or two hundred choices, I could predict the next four hundred with ninety percent accuracy, okay? So the statement 'I am sure no one can predict what I'm going to do' is wrong."

How does our mind create value and attach desirability to one particular decision as opposed to another? This is where the science becomes a little fuzzier, but, at the same time, much more interesting in terms of its implications for wisdom. It is Glimcher's view that value is ultimately determined in the medial prefrontal cortex (PFC); this is the inner (medial) portion of the more evolutionarily recent cortical part of the brain. But the PFC is kind of a basin into which all sorts of information about value is funneled from a number of neural subsystems. The exact number remains unclear—Peter Dayan's group at University College London has recently identified four separate subsystems in monkeys, and that doesn't include semantic (language-based) and symbolic evaluations that would undoubtedly play a role in human discernment. Glimcher likens these to "different modules, basically, that are helping to establish value that funnel into a final common pathway." The 2007 experiment provided some of the best evidence that all these evaluating substations, in fact, funnel into the prefrontal cortex.

Now that experiment, of course, posed a very simple problem involving financial evaluation, but in terms of more global neural activity, try substituting words like *discernment* or *reflection* when Glimcher speaks of "valuation." "The way I think about it is, valuation is a very complicated problem," he conceded. "It depends on emotional state, as it should. It depends on habits. It depends on recent trial-and-error experience. It may depend on symbolic knowledge, like percentages. But at some point, you have to bring all that information together. Because as John von Neumann noticed, as people have noticed for a long time, and philosophers noticed long before that, in order to choose, you have to lay out your options on a common scale. There's no way to choose between apples and oranges unless there's a common scale. And that really means at some level that to do efficient choice, you have to converge all this data into a final common scale and pass that to the decision maker. And the argument we've made is that the medial prefrontal cortex and perhaps the ventral striatum *are* that final common pathway."

In a sense, Glimcher is already at least half right about free will; our judgments are inevitable, indeed almost automatic, by the time our subjective calculation of value reaches the neural decision point, at which time the "most rewarding" option is automatically selected. But how do we decide what is "most rewarding"?

The very question suggests, in a funny way, that most of the action in terms of wisdom and judgment probably happens upstream of the decision maker. It lies in all the neural spigots that feed information and data into valuation—not just food, water, sex, and money, but memory, experience, culture (both knowledge and imitation), emotion, context, sense of time (and future rewards), fear of death, social altruism, and on and on. And while it might seem odd that a blunt motivational apparatus like the dopamine system might underlie the most abstract and ethereal of human virtues like wisdom or courage, traits that by their exalted nature would seem to deserve the neural equivalent of copper plumbing, the likely reality is that evolution's priority in building a good brain did not have to do with wisdom per se, only with how wise behavior might enhance survival. As Glimcher puts it, "Evolution, we have to believe, provided a very strong pressure for animals to do the smart thing, and that is to find a final common path, a common valuation, and make decisions based on it." The alternative, he said, is that the brain would become paralyzed by these conflicts in choice and essentially do nothing—except perhaps starve to death while trying to decide whether to eat an apple or an orange.

These experiments raise some intriguing noneconomic questions. If there is a single valuation and reward circuitry, as Glimcher's work implies, is there any feedback loop by which we might derive a reward from making a decision or taking an action that *feels* wise? And could this paradoxically selfless self-satisfaction be piggybacked on a neural system fundamentally designed to service basic needs, like keeping us quenched and nourished? Put another way, did someone like Gandhi learn to give himself a spritz of dopamine when he undertook a selfless act?

In recent years, neuroscientists have begun to study the impatience with which humans deal with thirst or hunger. If you haven't had a drink of water in a while, you're much more likely to want a small sip right away than to wait a little bit for a larger sip. Similarly, if you're starving, you're likelier to want a small snack immediately than to wait a little bit for a more substantial snack. That urgency to satisfy physiological needs—that *impatience*, Glimcher argues—is hard-wired into the brain, closely tied to respective sensors for blood glucose and hydration, and so evolution, not surprisingly, has attached greater neural value to the immediate gratification of these needs. Glimcher suspects that our basic valuation of sex is probably the same: the sooner, the better. But no one has yet done the experiment that would knit all these basic and abstract rewards together—that is, compare the neural valuation of food, water, sex, and money in the same experiment with the same individuals to see if the same brain circuitry is used. Glimcher, however, suspects humans use virtually the same neural valuation circuit for all of them, and his lab is now testing that hypothesis in an experiment on water, food, and money.

It should be obvious to any discerning reader that it is dangerous to extrapolate from a highly circumscribed exercise in economic decision making, such as the recent experiments at NYU and elsewhere, to something as grand and elusive as wisdom. As Glimcher himself is quick to point out, a "quagmire" awaits anyone who indiscriminately draws links between, for example, psychological explanations of behavior and neurobiological explanations. But let's put on our boots and wade into that bog for a moment.

First off, let's concede that experiments in neuroeconomics fail to capture the *gestalt* of wisdom in a number of ways. In the NYU experiment, subjects were asked to choose between an immediate or a delayed monetary award in *six seconds;* the life decisions we associate with wisdom obviously entail choices more complex, subtle, and uncertain than grabbing twenty dollars now or forty dollars in two weeks, and they usually require a much more elaborate process of deliberation and reflection.

In economic questions, there is often an obviously "right" answer (at least from the point of view of optimal economic behavior); in questions that challenge our wisdom, there is hardly ever a right answer in the same sense. What makes wise decisions so hard is that they must be taken in the murk of uncertainty and ambiguity, thereby clouding our ability to make clear evaluations; indeed, the real value of our choices sometimes doesn't become clear for years.

More important, in experiments where subjects are asked to choose between immediate and delayed rewards, an individual's sense of time will obviously inform her or his choices. Posing such questions to an experimental cohort whose average age is twenty-one might easily yield a different answer, and perhaps even a differently weighted neural process of valuation, than, say, asking forty-five-year-olds or sixty-five-year-olds.

Indeed, for me, one of the most profound (and largely unremarked) findings of the NYU experiment was how idiosyncratically each subject valued delayed gratification. Each of them, even though of essentially the same age and background, possessed—at the level of measurable excitability in a specific region of the brain—a different thermostat for impatience, a different neurobiological taste for reward, a different way of gauging what has the greatest value (which, in a philosopher's vocabulary, is another way of saying what is most important).

Where does that difference come from? What informs it? Is it purely biological, with genes shaping the excitability of the medial prefrontal cortex? Is it developmental, where early life experiences somehow tune that tissue? Is it cultural, where things learned or taught or merely observed (in terms of cultural impact on everyday economic behavior, is there a more persuasive form of imitation than keeping up with the Joneses?) somehow modulate that excitability?

No one knows, although as even Glimcher suggested earlier, people's financial preferences can be repaired by education. "If you explain it to people," he said, "they learn to get it right." So there is always room, even in a hardwired neural system, to tinker with the settings. In fact, at the time we spoke, he said financial traders on Wall Street had begun to explore meditation exercises as a way of optimizing their decision making ("The Street really gets this," as Glimcher put it at the time, although one suspects those on "the Street" might find meditation more useful now as a way of regulating negative emotion).

Some dearly held notions of what constitutes wisdom are directly threatened by this new research—not just free will but also conscious choice, the role of intuition, and the value of a deep, penetrating focus when contemplating a problem. A large body of work suggests that our brains reach decisions well before we're aware that a decision has been made, which raises the awkward notion that free will, such as it is, may be largely unconscious. John-Dylan Haynes and his colleagues at the Max Planck Institute for Human Cognitive and Brain Sciences in Leipzig, Germany, have shown in several recent studies that the brain races ahead—at least ten seconds ahead, which is a neural eternity—of consciousness in arriving at a choice. These choices are, admittedly, pretty simple, but the research challenges the notion that deliberation is better than intuition (although it doesn't say much about the role of deliberation in establishing subjective value in the first place).

Haynes believes that no good decision comes out of a knowledge vacuum. "You always have to know something about the domain," he says. "You can't intuit a useful decision out of the blue without having the relevant information." But he also points out that some studies have shown that, when faced with a decision or problem involving a large number of uncertainties, intuition may be better than reflection. "There are studies suggesting that conscious deliberation is good when the decision should be made on a few aspects, but is bad when you have to integrate information across a large number of details, whereas intuition offers you a more holistic version of processing." Nonetheless, most of the experiments reported to date involve (I hesitate even to say it) mindlessly simple choices, so these are still early days. When it comes to something more complicated, like choosing one's lifetime profession or even just buying a car, Haynes agrees that the jury is still out. "We still need to see," he said.

In some labs, the jury has already come back on complex decisions like buying a car. The Dutch psychologist Ap Dijksterhuis has almost single-handedly demolished traditional notions of deliberation with a series of experiments over the last decade showing that big decisions (like buying a house or car) are best made through unconscious thought. This intuitive, noncontemplative process—he calls it "deliberation-without-attention"—consistently results in greater satisfaction on the part of people who make decisions this way, at least in consumer choices. As he put it in a recent paper with Loran Nordgren, "People tend to engage in a great deal of conscious thought when they deal with complex problems, whereas they should engage more in unconscious thought." And the key to unconscious thinking, according to Dijksterhuis, is "thought without attention."

Another dearly held component of decision making is focused attention. William James devoted a good deal of attention to attention in his book *The Principles of Psychology,* and as Christine Rosen pointed out in a recent essay, James believed "steady attention was thus the default condition of a mature mind, an ordinary state undone only by perturbation." James himself wrote, "The faculty of voluntarily bringing back a wandering attention, over and over again, is the very root of judgment, character, and will."

And yet, paradoxically, there is such a thing as too much focus; neuroscientists know it as the "attentional blink." This neural tic works like this: If you are intensely focused on paying attention to a particular detail or picking a specific detail out of a random pattern of information, you neurologically cling so fiercely to the initial perception of it that you tend to miss seeing another example of it if it follows soon after. Many experiments have documented this momentary lapse of focus. Hence, the "attentional blink."

Some experiments now suggest that optimal attention requires a middle way between intense focus and plain old mind wandering. Richard Davidson, a scientist we'll learn more about in the chapter on compassion, has had a longtime interest in meditation and its potential to alter brain function. In 2007, his group published the results of an unusual experiment showing how meditation could "cure" the attentional blink. Volunteer subjects spent three months learning Vipassana meditation, which trains practitioners to, as one commentator put it, "cultivate awareness of stimuli without judgments or affective responses to those stimuli." In other words, a disengaged kind of attention. This well-controlled study showed that meditators learned to perform better on a standard attentional blink test than did two other groups, including some novice meditators. In an odd way, meditation seemed to allow a more detached, less effortful form of attention. If, as James suggested, attention is at the very heart of judgment, of determining what is important, it is nonetheless much more complicated a part of deliberation than our popular admiration for a steely focus.

. . .

There may be a tincture of wisdom in choosing forty dollars in two months rather than twenty dollars now, but does this have anything to do with the pragmatic life choices that psychologists like Baltes and Clayton thought about? Can it shed light on something as common as deciding which college to attend?

Glimcher's short answer is yes, and he believes it has everything to do with the difficulty of the decision. If you are choosing between Podunk Valley Junior College and Princeton, for example, the choice is fairly straightforward. But if you must choose among Harvard or Yale or Princeton, how do you decide wisely? Glimcher says—and this is at the level of speculation—that we use the same decision-making machinery but have to edit the knowledge or information we feed into it.

If you're interested in becoming a historian, for example, that will guide your knowledge gathering, which in turn will help you discern the relative strengths of the history departments at the three universities. So narrowing, or "editing," your focus is a preliminary (and sometimes intuitive) decision about what is important, and this almost inevitably recalibrates value before you make the ultimate decision. Again, this editing process—this neural filter of value, if you will—lies upstream of decision making. Hence, knowing how to edit down choices seems crucial to wisdom, even if the ultimate decision relies on the same workhorse machinery in the brain that chooses between twenty dollars now or forty dollars later. "Is there a difference in the neural architecture for hard decisions?" Glimcher asked. "Or is it just that decision time is a function of difficulty, and as the decisions get harder and harder, time goes up? It's the same machine. It just takes longer to converge on a proper solution."

It takes considerably longer to converge on a choice when we are confronted with too many options, and experiments have shown that we are terrible decision makers when we don't edit wisely—and most people don't. In a literally delicious set of experiments, Sheena Iyengar at Columbia University and Mark R. Lepper at Stanford University concocted a classic dilemma in decision making in which they asked participants in a study to choose their favorite flavor of gourmet jelly. When the options were limited to about six choices, people converged fairly quickly on their choice; when confronted with twenty-four or thirty different kinds of jelly, however, they took longer, often didn't decide, and, if they did, often expressed more regret about their decision.

Glimcher believes this boils down to a neural version of fractions and math, where the value we attach to any particular flavor is divided by the number of alternatives. When there are lots of choices, the "denominator" of decision making is large, so that the relative value of each jelly in your mind is subsequently smaller, reduced to the level of neural noise. As a result, you can't decide. "You sort of order the most important attributes for yourself, and use these as a way to edit down the set of objects that are being chosen between," he said. "And once you get that group small enough, you can choose."

The idea that a multiplicity of potential choices can be cognitively paralyzing actually has a striking parallel in clinical medicine. Neurologists have described an unusual state of paralysis in people with Parkinson's disease, which, if more broadly applied to the great unwashed horde of consumers, might explain why so many of us feel brain-addled when we go shopping. This form of paralysis does not affect muscles or limbs; it reaches right into the brain and affects our ability to choose. The clinical term is *decision paralysis,* and it arises—and has been demonstrated in human experiments—when people are confronted with too many possible solutions to a problem. The fascinating thing about this line of research is that when people are initially presented with a narrow range of options—say two or three choices—they have no problem reaching a decision. When the same two or three options are mixed in with a larger array of possible choices, however, paralysis sets in and some people are incapable of identifying and choosing the same option that appeared desirable earlier. After a certain threshold of complexity, it seems, deliberation is not helpful, but, rather, debilitating. And, it should be pointed out, Parkinson's is characterized by a loss of dopamine-secreting neurons deep in the brain.

The flaw, the loose thread in the rug of neuroeconomics, may lie in the definition of value (or preference), because most classical economists assume personal preference is unchanging and correct. "The decision-making module *has* to choose amongst objects that have values associated with them," Glimcher said, "and of course all the real action of decision making is setting up and either stabilizing or changing those values."

There is heresy buried in that mild statement, however. Neoclassical economists don't believe those values should change—ever. They don't believe, to put it in more general terms, that we should reevaluate our preferences, which are essentially our meta-decisions about what is important. Glimcher sees this editing process, the winnowing of options according to values, as a way to use the limited decision-making capacity of the brain to deal with difficult choices. "I think that's the only way an economist could really see this stuff," he said. "The only other way is that your preferences change, and that's more troubling, because it means people don't have stable desires. And for an economist," he added with an ironic laugh, "down that road lies only darkness."

As Montaigne surmised, something tells us when we've made a wise choice. It goes beyond self-satisfaction, and often beyond the immediate moment; we can also speculate that it increasingly satisfies relational goals rather than purely economic ones. Some of the recent neuroeconomic findings, preliminary though they may be, explicitly challenge "conventional wisdom" about what makes for keen discernment, sound judgment, and good decision making.

But it would be a mistake to dismiss the growing body of evidence about how the brain works, and what that means to our understanding of human traits like patience, altruism, moral judgment, attentional focus, emotional calm, other-centeredness—all the qualities we'll be discussing as pillars of wisdom. And it would also be a mistake to forget that everyone's interpretation of this new information is shaped by her or his intellectual, cultural, and emotional vantage point. And this gets back to the tension to which Paul Glimcher confessed when we first spoke. Like many of the other neuroscientists whom I visited, he did his obligatory squirming when asked about the intersection of neuroscience and wisdom. But his discomfort is both edifying and revealing.

"The economist in me is much less comfortable with these issues," he said. "You know, Friedman told us sixty years ago that a person's preference is a person's preference is a person's preference, and who are you to screw with that? There is no such thing as being too impatient. There's just the level of impatience you show. And what we want to do is maximize everyone's welfare without telling them what their preference *ought* to be. . . . This is, of course, at variance with some traditional psychological notions. People with high IQs tend to be more patient. Now the psychologist says, 'Well, that sounds pretty wise.' But the economist says, '*What?* Who are you to judge this group of people as more wise than that group of people? Both are getting what they want out of life. One wants immediate gratification, one wants delayed gratification, and who are you to say which is better off?' "

Glimcher paused to sigh, and then added, "So I find myself at war with myself over this issue. Especially as someone who obviously is good at delayed gratification, or I wouldn't be a successful scientist. But I also hear Friedman barking at me, and I think he's at some level right, because it's got a nice libertarian, relativist feel to it."

If honoring one's preferences—knowing exactly what one wants and then getting it, to put it bluntly—is the bottom-line value, then this neuroeconomic version of *Homo economicus* does bear a surprising resemblance to some of the classic definitions of wisdom: He or she is undoubtedly living the best life possible, is maximizing his or her choices, and for that reason is probably happy. And yet paradoxically, in a chapter about judgment, it is hard not to notice that many economists feel intense discomfort at being judgmental about human behavior; they are about as un-Socratic as you can be.

"As an economist, I really really don't want to stack people up, order people, based on their preferences," Glimcher admitted at one point. "It just seems wrong to me. The only thing as an economist I could really do in good conscience is order people by their internal consistency. So a first-order answer is, someone who is consistent but impatient, and someone who is hugely inconsistent but patient? I would have to go for the impatient guy is wiser, because he's actually maximizing his welfare better; he's getting what he wants more, via the nature of his consistent choice."

So that is at least one modern, neuroscientific version of wisdom: the consistency of your choices in getting what you want. This certainly aligns with some traditional notions of a good—and wise—life, and it's backed by a wealth of data about what guides our decisions. Still, it doesn't quite seem selfless or "paradigmatic" in the way Karl Jaspers saw

Socrates or Confucius or the Buddha. To quote the famous cognitive psychologist Peggy Lee, "Is that all there is?"

This basic neural machinery leaves us with a paradox worthy of wisdom. As Read Montague noted in a recent essay about free will, "The problem of choice for biologically evolved creatures is exactly an economic problem," where a decision is based on a comparison of the *perceived* value of possible choices. And as even economists have conceded, the kind of person who maximizes those rewards with the brisk efficiency of a calculator—the famous *Homo economicus*—is happy, yet also by definition selfish. So what happens when our selfish campaign for happiness clashes, as it so often does, with the goals of others? That is where wisdom, although still rooted in biology, enters the realm of *social* neuroscience.

CHAPTER SIX

MORAL REASONING

The Biology of Judging Right from Wrong

*We speak not strictly and philosophically when
we talk of the combat of passion and of reason.
Reason is, and ought only to be the slave of the
passions, and can never pretend to any other
office than to serve and obey them.*

—David Hume, *A Treatise of Human Nature*

THOSE SHREWD ANONYMOUS SCRIVENERS who compiled the Old
Testament wasted no time getting to sin. In chapter 3 of Genesis, God
draws a moral line in Earth's still-pristine soil beyond which mortals
should not transgress; and, of course, almost the inaugural act in the
annals of human behavior, after blinking our eyes and looking around, is
disobedience. Once Eve and then Adam ate the forbidden fruit, "the eyes
of the two were opened up, and they knew they were naked." Then, with
fig leaf and loincloth, they committed humankind's first cover-up.

Quickly surmising the cause of their shame, God thundered, "From
the tree I commanded you not to eat have you eaten?" On behalf of all of
us (but without really having gotten our consent), Adam and Eve implicitly admit their original sin by slinking out of the Garden of Eden. And,
like a deleterious mutation that entered humanity's moral germ line
from its inception just east of Eden, all of us are said to have inherited
shame (which is a psychologically, if not religiously, suspect proposition)
and, by extension, an intuitive knowledge of the difference between
right and wrong (which, surprisingly, looks more and more to be biologically true). The authors of Genesis put the lure of temptation in the
mouth of a creature with a forked tongue, but the serpent's words—

"Snakes were a symbol in the ancient world of wisdom, fertility, and immortality," according to biblical scholars—might also be heard as a bittersweet gift: "you will become as gods knowing good and evil."

An irreverent secularist might find two contrarian lessons about wisdom in this foundational myth. The fact that Eve took the first bite might suggest that women, from time immemorial, have always been a step ahead of men in using the intuitive power of their "opened" eyes to discern the murky currents of human nature. And if expulsion from the Garden of Eden was the price to pay for keener insight into the vagaries of human behavior, it might not have been a bad deal for humanity in the long run. By this reckoning, exile also marks the biblical origin, albeit painful and conflicted, of human wisdom. The more traditional interpretation, of course, is that the transgression in the Garden, chronicled at least as early as the sixth century B.C.—again, not long after the birth of the Axial Age—endowed humans with a primordial notion of moral judgment (and moral self-perception) that has legitimized centuries of religious acculturation of right and wrong.

And yet this rich, deep, and meaningful framework for morality has been put at profound risk by a mere two decades of neuroscience research, which threatens to overturn the whole Edenic applecart. What if moral judgment, so central a notion to all schools of philosophy and the centerpiece of every major religion, is not the conscious, deliberate, reasoned discernment of right or wrong we've all been led to believe, but is, rather, a subterranean biological reckoning, fed by an underwater spring of hidden emotion, mischievously tickled and swayed by extraneous feelings like disgust, virtually beyond the touch of what we customarily think of as conscience? What if Plato, Socrates, and Aristotle were nothing but a bunch of two-bit, fork-tongued, post hoc rationalizers? What if, every time we try to decide what is the "right" or "good" thing to do, we are merely responding, like dogs, to the otherwise inaudible whistling of the emotional brain? That is where moral philosophy is headed these days, and it's being driven by a new generation of philosophers and social psychologists, who have adopted the uniform of the lab coat.

Before there was neuroscience, before there were moral philosophers like Kant and Hume, before there was even a New Testament and an old Yahwistic tradition, the importance of goodness—of doing right instead of wrong—had already assumed central importance in the most influential thought systems in history. Confucianism asserts the primacy of *gen,* or goodness, in governing thought *and* action; "Not to act when justice commands is cowardice," Confucius said. "Set your heart on doing good," the Buddha urged. "Do it over and over again, and you will be filled with joy." And the entire edifice of Aristotelian wisdom rests on the clear discernment of right and wrong in the conduct of both private and public life. It is, Aristotle wrote, "impossible to be practically wise without being good."

But neuroscientists have recently tunneled underneath all that lofty rhetoric by philosophers and theologians, and they have begun to discover that judging right from wrong, and making decisions that wisely discern between the two, can also be glimpsed in the activity of the brain. In a sense, the path blazed in shame by Adam and Eve as they skulked out of the Garden of Eden has led, in a meandering fashion, to discrete neural areas in the brain like the anterior cingulate cortex, the insula, and the prefrontal cortex. These structures, remarkably, light up when we mull conflicts over what is the right, or moral, thing to do.

There is more at stake here than cultural and religious custodianship of right and wrong; the very act of moral reasoning is in play. Marc Hauser, a biologist at Harvard University and one of the leading researchers in this new field, writes in his recent book *Moral Minds* that "moral judgments are mediated by an unconscious process, a hidden moral grammar that evaluates the causes and consequences of our own and others' actions." He goes on to argue that this moral grammar is innate and hard-wired in our brains, deeply embedded in the emotional circuitry; as a result, Hauser says, "Our moral instincts are immune to the explicitly articulated commandments handed down by religions and governments. Sometimes our moral intuitions will converge with those that culture spells out, and sometimes they will diverge."

In other words, if humans obey the command "Thou shalt not kill," it is more because of biological intuition than biblical injunction. Conscious thought—the satisfying illusion of deciding what is right or wrong—is just a rational spritz of whipped cream heaped atop a large scoop of emotional instinct.

Once we begin to think of moral philosophy as a biological trait, it is impossible to avoid extending the implications of this radical idea into prehistory, indeed into deep evolutionary history. As the paleoanthropologist Jean-Jacques Hublin suggests, one of the key events in the evolution of our species was the rise of a social order that established group norms and, as he put it, "punishes the cheaters." We'll get to punishment in a later chapter, which discusses altruism (an indispensable part of wisdom related to moral reasoning). The question here is: How do we recognize right and wrong? How do we know the difference between moral and immoral behavior? And how can we reconcile this most fundamental form of human discernment, which speaks to who we are as individuals, who we become as groups, and what we value as societies, with the notion that it is fundamentally unconscious and deeply biological? In the arena of wisdom, how do we discern the right and good thing to do versus the wrong and bad thing, which is surely the very foundation for any personal construct of wisdom that will guide behavior?

This is a topic of timeless philosophic dispute, but one that has begun to attract the attention of biologists. So let's take this fight in a different direction. Let's take it into the brain. And one of the quickest ways into the brain is through the nose and mouth.

"As the sensation of disgust primarily arises in connection with the act of eating or tasting, it is natural that its expression should consist chiefly in movements around the mouth," Darwin wrote in one of his last books, *The Expression of the Emotions in Man and Animals,* which meticulously documents the repertoire of emotional gestures shared by humans and many other primates. "But as disgust also causes annoyance, it is generally accompanied by a frown, and often by gestures as if to push away or guard oneself against the offensive object." In this typically effortless bit of biological omniscience, Darwin hinted at an astounding idea: that the facial contortions elicited by the smell of a rotting fish, for example, probably evolved as a reaction to poisonous or "bad" food and simply got redeployed to express our inner feelings about a rotten person or fishy behavior. More than a century later, a neuroscientific experiment reported in *Science* made precisely that point (the title of the accompanying commentary—"From Oral to Moral"—said it all).

And yet even Darwin failed to connect his own dots on this point. As he noted correctly in *The Descent of Man,* "Of all the differences between man and the lower animals, the moral sense or conscience is by far the most important." But he went on to make the following important distinction: "It is the most noble of all attributes of man, leading him without a moment's hesitation to risk his life for that of a fellow-creature; or after due deliberation, impelled simply by the deep feeling of right or duty, to sacrifice it in some great cause." We may have to give Darwin only partial credit on this point. "Without a moment's hesitation" looks to be an accurate estimate of the rapidity with which moral intuition operates. What is beginning to look more dubious, however, is the idea of "due deliberation" in moral judgment. And the "nobility" of the moral sense is taking on a bit of water, too.

Jonathan Haidt, a psychologist at the University of Virginia, has made a tidy little career challenging many of our comfortable notions about moral judgment—its nobility, its shrewdness, the gravity of its due deliberation—by connecting the Darwinian dots that link the high road of "moral sense" and the low road of mouth-contorting emotions like disgust. In a famous essay published in 2001 entitled "The Emotional Dog and Its Rational Tail," he argued that moral judgment derives from "ethical intuitionism" in the emotional brain and not nearly so much from rational "moral reasoning." "Moral intuition," he wrote, "is a kind of cognition, but it is not a kind of reasoning."

In an instantly classic (and deeply subversive) experiment conducted several years ago, Haidt and his collaborators asked participants in a study to rate their moral repugnance to a number of hypothetical scenarios, such as a friend lying on a job résumé or survivors of a plane crash contemplating whether to cannibalize an injured child. But there was a hidden card in the experiment: some participants filled out their questionnaires in a room containing foul-smelling garbage in a waste container, and others sat at a sticky, stained, and unkempt desk. The presence of these disgusting but morally irrelevant environmental cues nonetheless provoked greater moral outrage than in people who filled out their questionnaires in a clean, odorless environment. Aside from suggesting that air fresheners might facilitate dispassionate moral judgment, what does this mean?

An earlier experiment by Haidt and his colleague Thalia Wheatley demonstrated that moral judgment could even be swayed by hypnotic suggestions of disgust; participants in the study were hypnotized and trained to associate a neutral word like *often* or *take* with disgust. Later, when the word was used in vignettes that posed moral problems, these volunteers again shifted their moral gearboxes into high dudgeon. In Haidt's view, emotions like disgust do not merely tilt moral judgment; they wag the tail of moral decision making.

Disgust, in fact, may drive the whole process. In yet another Haidt experiment, subjects responded to the following incest scenario: a sister and brother, enjoying a summer trip through Europe together, decide one night to have sex with each other. The sister is taking birth-control pills, and the brother uses a condom. They are both said to enjoy having sex, but they agree not to do it again, and decide to keep it a secret. Then experimental subjects were asked, Was incest morally okay in these circumstances? Most people found the scenario objectionable, but couldn't articulate a good reason why—there is no possibility of offspring (because of contraception), no possibility of emotional harm (because both siblings feel closer after the experience), and no possibility of setting a bad community standard (because they keep it a secret). All the subjects were left with was an inarticulate sense that it was just plain wrong. Haidt distinguishes this process of after-the-fact moral rationalization from true moral reasoning.

It can hardly be a coincidence, to bring this briefly but explicitly into the orbit of wisdom, that Leon Kass, the University of Chicago moral philosopher, is perhaps best known for an essay entitled "The Wisdom of Repugnance," in which he argues that physical revulsion and disgust are the body's (and mind's) way of telling a person that an action—human cloning, in Kass's main example—is morally wrong. (Perhaps it's also not a coincidence that Kass has written at length, and quite beautifully, in *The Hungry Soul,* about the philosophy, psychology, and moral merits of the food we eat.)

Disgust, repugnance, and rotten fish may seem like slippery ground on which to embark on a journey toward wisdom, rather like setting off on an epic pilgrimage to enlightenment from a garbage dump. But in recent years, psychologists like Haidt and Hauser, legal scholars like John Mikhail at Georgetown University, lapsed economists like Joshua Greene at Harvard University, and moral philosophers like Kass have been rolling around in the muck of emotion, asking how gut feelings can inform decision making, moral judgment and, in a larger (but usually unenunciated) sense, wisdom. According to Kass and other "deontologi-

cal" moral philosophers, emotions like repugnance are wonderful teachers and guides, informing us at the unconscious but felt level about what
is right and what is wrong. Our gut feelings are somatic messages from
our inner Yoda. As Kass writes, "Repugnance may be the only voice left
that speaks up to defend the central core of our humanity. Shallow are
the souls that have forgotten how to shudder."

Yet that shuddering bile of moral disgust flows through a venerable
bit of animal neural plumbing, and even repugnance speaks, like the serpent, with a forked tongue. Recent brain-scanning experiments suggest,
by no means definitively but in a highly provocative way, that the detection of moral conflict, and the machinery of moral choice, is embedded
in the anatomy of the human brain. In the most simplistic interpretation, these experiments suggest that moral repugnance and disgust essentially slather emotion over clear thinking. Indeed, a series of very clever
experiments has exposed precisely how we allow emotions to guide—
sometimes sagely, and sometimes irrationally—our moral judgments,
and then how we concoct elaborate rationalizations after the fact to
make these judgments appear reasoned. In dismissing the "wisdom of
repugnance" argument, psychologist Steven Pinker has noted, "There
are, of course, good reasons to regulate human cloning, but the shudder
test is not one of them."

What does all this have to do with everyday wisdom? In her last published paper on the topic, Vivian Clayton made an important distinction
between knowledge and wisdom. Knowledge, she wrote, tells you how
to do something. Wisdom tells you whether you should do it or not.
Should is a word that, like a bridge (or, for that matter, a brain), must be
engineered to bear a lot of weight and yet be flexible enough to retain its
structural integrity when buffeted by unusual stresses. Not surprisingly,
the "should" or "ought" of moral behavior has attracted a lot of philosophic attention over the centuries. The Scottish philosopher David
Hume, who sometimes seems a solitary but stout tree resisting the wind
of the Enlightenment, perhaps best articulated the role of emotion in
moral judgment—that we make the moral judgments that we do
because they are informed by emotion. "Morals excite passions," he
famously noted, "and produce or prevent actions. Reason of itself is
utterly impotent in this particular. The rules of morality, therefore, are
not conclusions of our reason."

Philosophers have traditionally rendered this conflict as a debate of heart versus mind, emotion versus reason. But we have reached a point where, at least in terms of the anatomy of emotion, the twain have begun to meet: *is* is merging with *should* and *ought,* knowledge with emotion, moral intuition with moral judgment. The title of a recent paper in *Nature Reviews: Neuroscience* put it bluntly, "From Neural 'Is' to Moral 'Ought': What Are the Moral Implications of Neuroscientific Moral Psychology?" The fact that the question can even be asked reflects the audacity of modern science; the idea that it might be answered in the not too distant future may change some of our fundamental assumptions about moral wisdom.

How does the human brain process a moral dilemma and tell us what we should do? Some of the most interesting recent research on how we reach the "ought" of moral behavior has been spearheaded by Joshua Greene, who might be described as a would-be buttoned-down economist who crossed the street to the messier, less rational side of human behavior a few years ago and has never looked back.

Greene set out to study business at the University of Pennsylvania, where he began his undergraduate studies in 1992. But he kept getting nudged down an unexpected fork in the road, fascinated by an early exposure to experimental psychology. Indeed, within a year, he had switched universities (from Penn to Harvard) and focus (from business to philosophy). During summers, he conducted independent psychological research with Amartya Sen, the Nobel Prize–winning economist, became fascinated by the implications of Jonathan Haidt's research, and completed an undergraduate thesis on the way moral judgment could be influenced by psychological factors. By the time he began his graduate studies at Princeton, he had landed with both feet in the moral philosophy camp; by the time he emerged with a Ph.D., he had become a cognitive neuroscientist.

"I view science," he wrote at the time, "as offering a 'behind the scenes' look at human morality. Just as a well-researched biography can, depending on what it reveals, boost or deflate one's esteem for its subject, the scientific investigation of human morality can help us to understand the human moral nature, and in so doing change our opinion of it."

It was while Greene was hanging out in Green Hall, home of Princeton's psychology department, that he came up with the idea for a brain-scanning experiment that continues to ruffle and rattle the field of moral philosophy. Not to give away the store ahead of time, but when Greene wrote up the results and conclusions in his doctoral thesis (which he is turning into a book), he titled it "The Terrible, Horrible, No Good, Very Bad Truth About Morality and What to Do About It."

The road to this awful truth is not a road at all, but, rather, obliges us to use a more old-fashioned means of travel (two rails) and a hypothetical but obsolescent form of transport (the trolley) to get us into a lot of postmodern subconscious trouble. The "Trolley Problem" has by now become so entrenched a feature of neuroscientific discourse, and of popular accounts of it, that it's probably fair to say that both metal and mental fatigue with the idea has begun to set in. But this fascinating thought experiment—originally created in the 1980s by two female moral philosophers, Judith Jarvis Thomson and Philippa Foot, to probe the nature of ethical judgment—represents an example of the kind of clever experimental scenario that can shed surprising light on the internal workings of the moral brain.

In the 1980s, moral philosophers sought to understand how people reach moral decisions by posing a hypothetical situation in which experimental subjects had to choose between two unpleasant courses of action. In the Trolley Problem, a runaway trolley is out of control and about to run over a group of five people. The only thing that can be done is to flip a switch and shuttle the trolley onto a side track, where it will unfortunately hit and kill one person. If you were faced with this decision, would you do nothing (allowing the trolley to kill five people), or would you switch the trolley to the side track (knowing that this action would result in certain death for one person but save five others)? The general consensus among moral philosophers, and among the test subjects who have grappled with this dilemma, is that minimizing the loss of life by switching the trolley to the side track is an ethically defensible decision. Indeed, when Hauser and his Harvard colleague Fiery Cushman posed this question to hundreds of subjects in a series of Web-based variations on the Trolley Problem, they found that between 75 percent and 90 percent of respondents said pulling the switch (and killing the hapless laborer) was morally justified under the circumstances.

So far, so good. But an ethical wrinkle arises when researchers slightly alter the scenario. As in the previous example, the runaway trolley is bearing down on a group of five people. But this time you happen to be on a footbridge going over the track, standing next to a "large" (as the scientific papers politely describe him) "stranger" while the runaway trolley is streaking toward the group of five doomed people. You are told that the only way you can save the five people is by hurling the fat man off the bridge and into the path of the trolley, causing it to derail, thereby saving the other five people. The moral math remains the same: The death of one person would save five others. In this scenario, however, most people say pushing the fat man off the bridge is morally indefensible.

As a moral philosopher, Greene wondered what was going on in the heads of people who were willing to kill one person to save five people in one scenario but not willing to kill one person to save five in a slightly different situation. In fact, he wondered what their brains looked like, in real time, as they wrestled with this well-known moral quandary. So he and several collaborators at Princeton used the Trolley Problem and the footbridge variation as "probes" while scanning the brains of volunteers in an fMRI machine.

What they discovered is that moral decisions like this provoke significant conflict between emotional and cognitive parts of the brain. In particular, when someone was confronted with the morally unpleasant need to throw a fellow human being off a bridge—what Greene calls an "up close and personal" violation of appropriate moral behavior—she or he displayed heightened activity in core emotional areas of the brain; and among those who ultimately opted for the heave-ho, it took longer to reach that decision, as if it required more time for the cognitive, "rational" part of the brain to overrule the emotional part.

If you accept the utilitarian notion that sacrificing one life to save five is morally acceptable, Greene argues, then emotion can steer you toward an irrational decision to sacrifice five lives in order to spare one, no matter how right or good this decision "feels." Greene and his colleagues published these initial results in *Science* in 2001. While the paper didn't register much outside the field (perhaps because it was published three days after September 11), it caused quite a kerfuffle in moral philosophy circles. As Kwame Anthony Appiah slyly noted, "The philosophical commentary on these cases makes the Talmud look like Cliff[s] Notes, and is surely massive enough to stop any runaway trolley in its tracks."

Before getting to the general implications, I want to season the discussion here with more than a few grains of well-needed salt, because many of these caveats get lost in the inevitably abbreviated accounts we often read in the newspaper or online of provocative fMRI findings like this. First, the initial study was very small; only nine subjects participated in the 2001 *Science* paper (a follow-up paper in *Neuron* in 2004 reported data on forty-one people). A potentially more significant shortcoming—and one not unrelated to similar concerns about other fMRI findings that bear on wisdom—is that all the subjects were college undergraduates. This sample may represent a narrow, and perhaps morally callow, band of ethical judgment, especially because, as we've seen in other research, some fMRI studies suggest significant differences in emotional processing of information by younger versus older adults.

Finally, it's not unreasonable to wonder if there might be at least a little cognitive dissonance in experimental subjects as they are asked to perform these tasks, a kind of commonsense skepticism about a highly artificial and possibly unrealistic scenario. Could anyone be absolutely certain that the five people on the tracks were doomed? Might not some of them have heard the runaway trolley hurtling toward them and jumped out of the way? Could a large, or overweight, or even enormously obese, *Guinness World Records*–worthy Fat Man, if pushed off a footbridge, truly derail or stop a trolley? How would you know such a person would even land on the tracks? I'm five-six, a couch potato, and resigned to my spud status. Could I possibly heave a three-hundred-pound man, presumably in an advanced (not to say agitated) state of physical resistance, off a footstool, much less a footbridge? Even if I could, would my aim be true? Could I time the push just right to avert catastrophe on one try, or might I mistime the shove, inadvertently leading to the deaths of *six* people? And what if, in a fit of mortal pique, the fat man grabbed hold of my ankle as I shoved him off the footbridge ("Gotcha, you bastard!"), promising me an "up close and personal" encounter with the runaway trolley, and resulting in *seven* dead people? As my twelve-year-old daughter remarked when she looked at a cartoon of the Trolley Problem in *Science,* "This is so stupid. Why didn't somebody just yell?"

These probably seem like trivial objections, but they hint at precisely the messy variables that experimentalists, and sometimes philosophers, need to eliminate to keep a lab experiment or thought experiment neat and clean, even though most of us would acknowledge that real life is dirty, open-ended, bedeviled by complexity, and hardly ever neat and clean. If that's life on the other side of the philosophic tracks, it's also where most of us live, and where we're likelier to find wisdom. All of which is to say that we need to remember that these findings, as intriguing as they are, grow out of highly artificial experimental conditions that do not, cannot, and never will *fully* capture the complexity and contradictions of real-life moral dilemmas.

Having said that, however, these are important experiments that challenge us with a disturbing message about the way we think and arrive at moral judgments: that emotion, clearly an important component of wisdom, can both inform and *deform* moral judgment. As Greene has pointed out in numerous papers, including a witty and provocative 2009 essay entitled "The Secret Joke of Kant's Soul," Western moral philosophy over the last two hundred years has seen a pitched debate over how humans decide on the "right" course of action (surely a central element in any definition of wisdom). On one side are the utilitarians, characterized by Jeremy Bentham and John Stuart Mill in the eighteenth century and the Australian philosopher Peter Singer in ours, who have argued that moral judgment promotes (or should promote) the greatest good for the greatest number of people (a point we'll return to in the discussion of wisdom and political leadership).

On the other side are the so-called deontologists, who believe that individual rights and duties (including, by extension, the right to life of the fat man on the footbridge) take moral precedence over any greater good that may come from violating those rights. As Kant's foundational work, the *Critique of Pure Reason,* argues, these judgments are achieved by exhaustive, excruciatingly deep reflection and deliberation—the exercise of rational human thought, in short, at the highest possible level.

But then again, maybe not. As Greene and colleagues wrote in 2004, "We propose that the tension between the utilitarian and deontological perspectives in moral philosophy reflects a more fundamental tension arising from the structure of the human brain." By taking this fight into the brain (and therefore into the realm of biology), Greene also argues that there is an evolutionarily regressive component to our moral thinking. The gut-level, emotional reaction to unpleasant moral choices like tossing the fat man off the footbridge, he believes, reflects emotional responses inherited from our primate and prehistoric ancestors, where every social dilemma was "up close and personal."

These emotional decisions probably gave humans, in that earlier social and environmental context, a survival advantage. "In contrast," he and colleagues write, "the 'moral calculus' that defines utilitarianism is made possible by more recently evolved structures in the frontal lobes [of the brain] that support abstract thinking and high-level cognitive control." In other words, Greene speculates that the newer, rational part of the human brain must override occasionally conflicting moral evaluations from the emotional part of the brain in order to reach the kind of abstract decisions that promise the greatest benefit to the greatest number of people. Hence, the people who decided to throw the fat man off the footbridge in Greene's experiment required more time (in other words, they "thought" or deliberated longer) to make a decision and showed greater activity in the parts of the brain that typically exert cognitive control.

This idea has dropped like a grenade in the fields of philosophy, psychology, theology, and wisdom, and not without pushback from moral philosophers. Appiah, for example, points out that the Trolley Problem is a contrived "moral emergency," meaning a situation in which time is limited, courses of action are simple, something of great moral consequence is at stake, and you are uniquely positioned to intervene; those four features, he continues, "together make these situations very different from most of the decisions that face us." Moreover, he finds it reassuring that the idea of tossing the fat man causes pause. "Holding onto revulsion against killing people 'up close and personal' is one way to make sure we won't easily be tempted to kill people when it suits our interests," he writes. "So we may be glad that our gyri are jangled by the very idea of it."

On the other hand, the Greene experiment makes the "wisdom of repugnance" seem not especially wise (although it still stands as a tour de force of after-the-fact rationalization of what might be fundamentally bad emotional advice from the most ancient part of the human brain). If Greene and his fellow travelers in experimental philosophy are right, there is a delicious irony in the history of thought. The implication of the recent fMRI studies, Greene writes, is that Kant, the quintessential rationalist, was, "psychologically speaking, grounded not in principles of pure practical reason, but in a set of emotional responses that are subsequently rationalized."

How might this process happen in a mechanistic, neuroscientific way? In Greene's experiments, people confronted with the Trolley Problem—an *impersonal* moral dilemma, in his view—displayed greater activity in cognitive areas of the brain, like the dorsolateral prefrontal cortex; to grossly simplify, this seems to be an area that wheels into action during utilitarian deliberation. But during the footbridge variation, when the moral dilemma became "personal" and people were forced to contemplate heaving a fellow human into the path of an onrushing trolley, they showed heightened activity in parts of the brain known for emotional conflict and social cognitive processing, including the medial prefrontal cortex, the posterior cingulate/precuneus, and the superior temporal sulcus/temperoparietal junction.

To most laypeople, these are meaningless duchies in an evolving map of moral thinking, and rightly so; the borders, boundaries, and interstate commerce of this moral realm remain blurry at best at this early point in time. At another level, however, we may be holding in our trembling hands an out-of-focus but breathtaking action snapshot of moral thinking on the fly.

Because this is frontier science, the neural map of moral judgment is inevitably contested territory, too.

Antonio Damasio of the University of Southern California, who has historically used the dark gift of human brain damage to shed brilliant light on behavior, has found that people who have suffered injuries to a different part of the brain, the ventromedial prefrontal cortex, tend to display abnormal moral judgment; in fact, they tend to be pathologically utilitarian, making "impersonal" moral judgments in situations that normally provoke emotional conflict and elicit emotional processing. Jorge Moll, a Brazilian neuroscientist, and his colleagues have published recent findings that similarly pinpoint "moral" activity in the ventrome-dial PFC; Moll believes this region, unlike the ones identified by Greene, reacts to moral challenges that have specific sociomoral content, dilem-mas that engage both guilt and compassion. And in a recent study in *Science,* researchers at the University of Illinois and the California Institute of Technology showed that the brain works hard to assess the moral equity or inequity of a situation, but then passes along this evaluation to another part of the brain, one that seems to specialize in making the most efficient choices. Perhaps the most important point about all these studies is that they were published within the last several years, with the clear implication that this is a wide-open, rip-roaring, and unsettled area of neuroscience. It will take at least ten years for the dust to settle, according to Walter Sinnott-Armstrong, a moral philosopher at Dart-mouth, who already sees evidence that there will not be one, all-purpose brain circuit for moral judgment. "In the last couple of years," he said, "new papers have distinguished different circuits for different aspects of moral judgment. The trend is going in the direction of different moral judgments being processed in different parts of the brain. It's important for wisdom if you tie this to the ancient Greek tradition of claiming the unity of all virtues, and that you can't have one without having all of them. But it looks like people can have parts of moral wisdom without having other parts of moral wisdom." Moreover, all this "disunity of morality" resides upstream from any decisions and actions we might take based on moral perception. "None of these scientific studies is about decision making," Sinnott-Armstrong says. "They are about moral *judgment.*"

Nobody wants to be a fat man, according to the old Jethro Tull song, but there may indeed be something rational about our reluctance to pitch him overboard. Bioethicist Thomas H. Murray, president of the Hastings Center, points out that placing greater value on one life close at hand than on five in the bush can indeed be considered utterly rational if it reflects our experience of having "special" emotional relationships with people—a relative, a child, a dear friend, some exceptional form of social affiliation. Perhaps our brains are built to calculate utilitarian value when we are faced with an abstract moral or social dilemma that affects the smooth functioning of the entire group, while at the same time retaining the capacity to assert a more primal, emotional moral judgment when the dilemma poses emotional exceptions to utilitarian deliberation, such as the involvement of kin or "special" people. "Morally speaking," Murray says, "the relationships that we have are sources of great responsibility, but also sources of our great flourishing. Special relationships are not accidental, but essential elements of being a human being, a moral human being, and a flourishing human being."

This barometric reading of moral judgment, with its fluctuations of socially sensitive pressure, seems consistent with wisdom to me. Indeed, the oscillation between the two moral impulses, depending as they would on real-world social contingencies and context, would require precisely the kind of judgmental tussle that a serious moral problem would demand, as well as the contextual flexibility to which the wise mind is always attuned. It's important to note that Josh Greene does not argue that emotion has no role in sound moral decision making. Rather, in pitch-perfect accord with more traditional thinking about wisdom, he believes the true conflict occurs in knowing when to trust it and when to discard it. There were, as he suggests, undoubtedly situations in our distant evolutionary past in which an instantaneous emotional judgment in an "up close and personal" situation was life-preserving. That part is hard-wired into the survival circuitry.

It may strike us as odd and perhaps dehumanizing to think of moral judgment, and the wisdom that comes along with it, as nothing more than the hiss and pop of a few esoteric brain circuits. To use a word that seems unfortunately burdened these days with exceptional cultural and ideological baggage, there doesn't seem to be much *dignity* in deciding right from wrong in such a cravenly biological, neurally furtive, and sometimes literally unthinking manner. Jonathan Haidt often uses the metaphor of the elephant to convey the role of emotion in human cognition and moral judgment—a big, shaggy, essentially unmanageable beast over which we exercise virtually no control and upon which we cling for dear life as it rumbles along. To extend the metaphor, we are not agents of moral judgment, merely passengers on a powerful animal as it crashes through thickets of moral dilemmas, and our reasoning is literally an afterthought. "People who claim emotion deforms judgment?" Haidt says. "Well, look at reasoning. We're *terrible* at it."

But Greene sees more potential for balance between the two. At some level, the rational part of our brains must be getting through, too, because we *are* capable of moral judgment, and of utilitarian thinking, and of an altruism that rises above pure emotional self-interest. Just as it makes biological sense to build automaticity into emotional reactions (the fear of snakes, not to mention talking and tempting serpents, clearly enhances survival), it also makes sense to add emergent qualities like true moral reasoning and wisdom to temper emotions, especially when automatic emotional judgments—anger, disgust, moral outrage—have obvious social consequences. It is not easy to exert cognitive control over this beast, and it often fails, but the neural plumbing provisionally sketched out so far at least allows a role for higher-order, top-down, thoughtful, deliberative input, as well as the bottom-up, rumbling, stumbling, elephantine crash of emotion.

So when it comes to moral judgment, maybe wisdom's role here is as a kind of "elephant whisperer," rationing bits of reason in a way that soothes and nudges and perhaps even steers our inner pachyderm, allowing our best moral instincts to emerge. Confucius always understood that you could not have goodness without wisdom and courage. The emerging biology of moral choice challenges us to summon enough wisdom to discern when emotion distorts our moral judgment, and enough courage and strength to deflect the compass needle of the passions and point it toward loftier, larger, more meaningful goals. The sheer difficulty of that challenge may alone explain why so many of us find it so hard to be even a little wise.

CHAPTER SEVEN

COMPASSION

The Biology of Loving-Kindness and Empathy

When I was young, I admired clever people.
Now that I am old, I admire kind people.

—Abraham Heschel

It would not be true to say that the cultivation of
loving kindness and compassion is part of our
practice. It would be true to say that the cultiva-
tion of loving kindness and compassion is all of
our practice.

—Buddha

IN THE MIDDLE OF THE TWELFTH CENTURY, during a brutal civil war whose very name barely rises above the mists of medieval minutiae, the army of Emperor Conrad III laid siege to the town of Weinsberg, in what is now southern Germany, eventually forcing its citizens and their leader, a duke named Guelph, into capitulation. The endgame to such a bitter confrontation usually entailed pillage, rape, destruction, and the obligatory execution of every adult male on the losing side, including (and especially) its leader. The fate of women and children depended on the discretion of the victorious commander (or, more to the point, on the degree to which he could control his vengeful soldiers).

When Conrad III finally vanquished his foe in December of 1140, he sent word that the women of Weinsberg would receive safe conduct to vacate the town, but under one condition: They had to leave all their worldly possessions behind, save what they could carry in their arms.

When the appointed hour for the evacuation arrived, according to medieval chroniclers, the women of Weinsberg staggered out of the besieged city, bearing on their shoulders their most precious worldly possessions—their husbands, including the defeated leader Guelph himself.

"The Emperor took such great pleasure in the nobility of their courage that he wept with delight and wholly subdued the bitter and deadly hatred which he had borne against this duke," Montaigne later wrote, "and from that time forward treated him and his humanely." The great French essayist went on to confess that he himself was "wonderfully lax in the direction of mercy and gentleness. As a matter of fact, I believe I should be likely to surrender more naturally to compassion than to esteem."

Many traditional working definitions of wisdom, whether of philosophical or psychological inflection, reserve a special role for compassion. By compassion is meant not only the willingness to share another person's pain and suffering; in a larger sense, it refers to a transcendent ability to step outside the moat of one's own self-interest to understand the point of view of another; in a still larger sense, it may take this "feeling for" to the level of mind reading, for the theory of mind—one of the most powerful implements that evolution placed in the human cognitive tool kit—requires us to understand the way another person's feelings inform his or her intentions and actions. Being attuned to the sufferings or struggles of another person (or, in the more expansive embrace of Buddhist compassion, another sentient being) ramifies into a broad array of human temperaments and behaviors: empathy, sympathy, charity, generosity, cooperation, and, in the broadest psychological sense, other-centeredness (as opposed to self-centeredness).

We can intuitively grasp the centrality of compassion and other-centeredness in human wisdom when we grab hold of the very long thread of the Golden Rule. Explicitly articulated by Confucius in the sixth century B.C. ("Never do to others what you would not like them to do to you"), this code of behavior anticipated by nearly six centuries the many famous acts of loving-kindness attributed to Jesus in the Bible, and it is now viewed as so fundamentally biological that neuroscientist Donald Pfaff, writing 2,600 years after Confucius, recently argued that "we are hardwired to follow the Golden Rule."

From this biological perspective, compassion would appear to represent one more bit of neural plumbing that connects emotion to thought and agency. Nonetheless, in platonic formulations of wisdom, reason is often invoked like a fire extinguisher to douse feelings like compassion. Indeed, a chilly draft of dispassion blows through the most famous antechamber in the history of philosophy, where Socrates spent his last hours on earth. It is not just that Socrates the philosopher equated death with an exhilarating liberation from emotions and feelings; it is that Socrates as husband and father, the wisest man in the world, shooed away his wife and young son long before the hemlock arrived, the better to converse more freely with his peers about immortality, the soul, and wisdom.

Few scenes in ancient history are more moving, yet it includes this strange moment of familial disconnect, a cold and emotionally dismissive episode that brings to mind Nietzsche's characterization of Socrates as "icy." We know Socrates' family was distraught, because Phaedo refers to his wife "crying hysterically" as she was led away; Socrates made no comment, and immediately began talking philosophy with his male cronies. To bring this full circle neurologically, twenty-four centuries later, the cries of a "distressed woman" have been piped into the ears of experimental subjects as they lie in an fMRI machine, so that scientists can try to identify the parts of the brain that become activated during feelings of compassion.

Is compassion an innate, biological instinct in humans, and thus worthy of scientific attention, or is it a culturally transmitted form of behavior best probed by developmental psychologists? Are there fixed quanta of compassion with which we are all born, or is it a quality that can be amplified and enriched through life experience and, perhaps, training? Until recently, it was pointless even to ask those questions. Few people viewed compassion (like wisdom) as a legitimate subject for empirical study, and fewer still believed it *could* be studied, at least in a rigorous and reductive way. As Anne Harrington, a historian of science at Harvard University, has pointed out with brutal candor, "the world that comes into view through the focusing lens of science is, at its deepest explanatory level, one in which compassion is irrelevant."

But compassion has always been the subtler, more submerged underside of human qualities that have traditionally attracted enormous amounts of research attention—notably, aggression and interpersonal violence. If the study of negative and antisocial emotion has long been considered legitimate, why not examine a clearly pro-social and positive emotion such as compassion? Compassion rounds and softens the hard-edged lessons of experience; compassion keeps the directional arrow of human agency pointed outward toward social interaction, as opposed to inward toward isolation and withdrawal; compassion warms human thought when intelligence becomes too impersonal and chilly; compassion motivates action in the direction of social good. "Knowledge without compassion is inhuman," said the particle physicist Victor Weisskopf. "Compassion without knowledge is ineffective." He might as well have been talking about wisdom, because it lies somewhere in between, or perhaps floats somewhere above, those two antipodal qualities.

The world has changed quite a bit since 1995, when Harrington wrote so despairingly of science's disinterest in compassion. Researchers have taken their first gingerly steps into the neurobiology of compassion and related emotions. These initial scientific studies of compassion belong to a growing movement in cognitive neuroscience that opts to focus not on pathology or diminished mental capacities, but, rather, on exemplary mental ability and expertise. These researchers have recruited excellent practitioners of one skill or another—cabdrivers for sense of direction and memory, or piano players for their motor coordination and "muscle memory"—to seek better understanding of extraordinary brain performance.

And that is why, since the mid-1990s, the road to a scientific understanding of compassion runs from the remote upland villages around Dharamsala in northern India, where an estimated 130,000 exiled Tibetan Buddhists live and practice, to a shiny neuroscience laboratory near the shores of Lake Mendota in Madison, Wisconsin, where the first neural traces of compassion have begun to appear in our twenty-first-century measuring machines. In the neural geography of wisdom, we have even managed to locate a few places in the brain associated with loving-kindness.

At exactly 8:48 a.m. on an unseasonably brisk, overcast June morning in Madison, with a hint of rain in the air, Matthieu Ricard climbed out of a Suburu Outback station wagon driven by his friend Richard J. Davidson (license plate: EMOTE), adjusted his saffron-and-purple robes over bare arms, and then sauntered into the Waisman Center of Neuroscience for an arduous day of meditation. Over the previous few days, Ricard, a French-born Buddhist monk who lives in a monastery in Katmandu, had flown from New Delhi to Newark, led a meditation workshop for financiers from a Manhattan stockbrokerage firm at the Harvard Club, discussed a new book project, and then traveled to Wisconsin, where he has participated in experiments since 2000. As he walked down a long corridor, he paused to hug old friends (he has come to Madison half a dozen times for these experiments) and greet people he'd never met before, including a bemused maintenance worker holding a paintbrush.

Within a few minutes, he had slipped out of his Top-Siders and settled his sixty-two-year-old frame onto a green pillow on the floor of a small, dark, nominally soundproofed chamber. Then scientists attached a hairnet of 128 electroencephalograph (EEG) electrodes to his shaven skull. Each electrode would measure electrical activity in this Buddhist brain as it meditated, capturing any changes in milliseconds, while nearby computer monitors would provide a running display of the readings. (Neuroscientists prefer EEG to MRI brain scans in certain experiments because the technique captures instantaneous changes much better.) "*C'est bien, Matthieu?*" asked Antoine Lutz, the French-born scientist leading the experiment.

"Okay," Ricard replied. With that, the heavy door was closed, and the experiment began.

"We have found a correlation between a deep state of meditation and changes in the EEG activity of the brain," explained Lutz, who published his first monk data in a 2004 article in the *Proceedings of the National Academy of Sciences,* a leading international journal. That study reported that expert meditators who generated a state of "unconditional loving-kindness and compassion" produced accompanying changes in the brain, especially something called "gamma-band oscillations" (about which more in a moment). Now they are gathering more data, and asking expert meditators like Ricard to estimate the intensity of their mental states (on a scale from one to nine) as their brain waves are being recorded.

Nearly an hour into the experiment, after segments devoted to "focused attention" and "open presence" meditation, the voice of researcher Andy Francis was piped into the little room. "The next one will be compassion," he said. According to the experimental plan, Ricard would spend the first minute in "neutral," not meditating, to provide a baseline of brain activity. After a pause of exactly sixty seconds, Francis leaned toward his microphone and said in a flat neutral voice, "Compassion, compassion."

At a scientific seminar in New York several months earlier, Richie Davidson had made the point that expert meditators could turn it on "like that," which he indicated by snapping his fingers, and although Ricard's compassion meditation was not as instantaneous as snapped fingers, it would probably strike most people as impressively quick. After fifteen seconds, he rated his meditative state at two; by thirty seconds, he was up to three; by thirty-nine seconds, he was up to four. He reached five at the one-minute mark, and achieved a lofty seven at two minutes. These are self-reported ratings, of course, and deserving of every bit of skepticism that self-reports inevitably invite. But since dedicating his life to the practice of Buddhism in 1972 (literally the day after he received a Ph.D. in molecular biology in the Paris laboratory of Nobel laureate François Jacob), Ricard has logged well over ten thousand hours of meditation, and I couldn't avoid the sense that I was watching a powerful neural engine rev up, roar, and accelerate like a race car, cognitively going from zero to sixty in the figurative snap of a finger.

As Ricard's meditative state deepened, the EEG readings scrawling across Andy Francis's computer screen visibly changed. At the beginning, they had been a series of thin, disorganized lines, the electrophysiological equivalent of lank, unkempt hair; by the time Ricard reached a self-assessed eight, about seven minutes into his compassion meditation, the EEG readings had an intensity of amplitude, looking like bushy eyebrows, and an overall coherence and synchronicity that were apparent even to my novice eye. "With a few practitioners," Francis confirmed, "you can see the change in the raw data, in real time, when you say, 'Meditate.' But that's very rare." At the risk of sounding credulous (if not downright corny), the electrical activity measured by dozens of electrodes plastered all over Ricard's noggin seemed in remarkable harmony. Then the muffled sound of a toilet flushing in a nearby bathroom could be heard, and whether this perturbed Ricard's meditative state or not, his subjective ratings began to drop.

Later on, in his office, Davidson agreed that the coherence and "synchrony" of these brain waves was so striking that it was obvious even to nonexperts. "It doesn't require any fancy computer interpretation," he said. "You can actually *see* it with the naked eye, it's so pronounced." His current thinking—"pure conjecture at this point," he conceded—is that the pattern of brain activity observed during meditation, known as a "gamma oscillation," might be a way for monks to coordinate activity in multiple parts of the brain while experiencing intense compassion. "What is striking about the gamma," Davidson said, "is the synchrony across brain regions that we see. From their own descriptions, it's kind of a panoramic awareness that's fused with compassion. But the elements of the panoramic awareness suggest that there are many different systems in the brain that become functionally integrated, which may be a neural correlate of this experience of a panoramic or vast awareness." A study by the Wisconsin group, published in 2008 in the journal *Public Library of Science One,* reported that fMRI studies of monks—and also novices who followed a two-week program of compassion meditation written by Ricard—showed increased activity during compassion meditation in several distinct parts of the brain. In particular, "sounds of distress" roused activity in the insula and anterior cingulate.

Ricard emerged from the chamber in Madison looking fresh and energetic after the morning's experiments. "You have to come to a lab to do meditation these days," he said with a chuckle, wiping his face with a towel.

Just before Ricard had stepped into the meditation chamber, I had brought up the issue of wisdom in passing, and when he emerged ninety minutes later, still wiping electrode gel off his bald pate, he immediately picked up the thread of the conversation. "I was thinking about this wisdom thing in the chamber," he said, "between the parts of the experiment." He went on to explain that, from the Buddhist perspective, there are two aspects of wisdom. One has to do with discerning the nature of reality, looking beyond the way things appear to consider their true inherent reality (this aligns nicely with ideas of discernment and knowledge that, in the good Socratic sense, make wisdom a truly investigatory exercise). The other has to do with using that knowledge or discernment as a motivation for action, with the specific aim of increasing happiness and diminishing suffering (this, in a sense, allows emotion to help guide action).

"Compassion is part of both," Ricard continued. "Compassion is based on understanding the interdependent nature of reality, and this leads to the recognition that I want to avoid suffering and be happy, that happiness and suffering are completely interconnected, so that leads to compassion." He added that "selfish happiness is completely at odds with reality, so it is not part of wisdom."

As an example, he cited a hypothetical case in which a person who accidentally falls overboard from a ship in the middle of the Pacific Ocean is visible from an airplane as he is about to drown. As the man thrashes in the water, the typical human reaction to his plight is one of pity, because there is nothing that can be done. "But if we know that there is an island nearby, which is hidden in the mist but could be reached by the person fallen at sea, then our reaction will be different," Ricard said. "We know there is definitely a possibility for the person in the ocean to survive, if only he can find the island. With this knowledge, our compassion will be much stronger and more intense. In a similar way, the fact that, through wisdom, one comes to understand that sentient beings can free themselves from suffering is a source of a much greater and powerful compassion than if we did not have the same kind of understanding and wisdom about the possibility of freedom, and about the means to achieve it." (As an aside, it is striking that while so much of modern neuroeconomics focuses on financial self-interest, much of traditional Buddhism focuses, as Ricard suggests, on a compassion-based emotional self-interest, where doing good in a pro-social way pays a different, emotional dividend in terms of personal happiness.)

The study of compassion and loving-kindness occupies, it need hardly be said, a place on the periphery of modern neuroscience. But it is inching its way closer to the mainstream, largely because of a decades-long effort by the guy who drives that car with the EMOTE license plate. Known to virtually all as "Richie," Richard Davidson still has the gangly, bright-eyed, tousle-haired look of the Brooklyn adolescent he used to be in the 1960s. He is now the Vilas Professor of Psychology and Psychiatry at the University of Wisconsin, and many of the biological issues he explores these days at the Laboratory for Affective Neuroscience—positive emotion, resilience, compassion, and emotional regulation—first began to pique his intellectual curiosity in the early 1970s, when he was a graduate student at Harvard University with an interest in, as he once delicately put it, "altered states of consciousness and meditation." If the term *altered states* evokes the sixties' pharmacologic variation on the "know thyself" theme, you're not far off; Davidson admits that some youthful dabbling in recreational experimentation helped inform his scientific curiosity.

As he explained in a semiautobiographical 2002 essay, he began to wonder if the mind could, in effect, be *trained* to increase one's well-being, and this led him to the fields of psychology and eventually neuroscience. Feeling, as he put it, "an acute lack of experiential learning in my formal graduate training," Davidson began to experiment in meditation himself. In the early 1970s, after finishing his second year of graduate school at Harvard, Davidson took an extended three-month trip to India and Sri Lanka, which included time at a retreat in northern India, where he underwent an intensive exposure to Buddhist mindfulness meditation. The experience was transformative.

One of the central goals of Buddhist practice, dating back to the initial wanderings of Prince Gautama, is to improve one's happiness. This goal of personal happiness does not intuitively align with the often bittersweet, other-centered, self-denying aspects of wisdom. But the Dalai Lama, the spiritual leader of Tibetan Buddhism, has argued that the Tibetan word for compassion, *tsewa,* does not necessarily imply that compassion is directed only outward toward others, but that it also can be directed inward, toward the self, so that it is consistent with the search for happiness. "In a way," the Dalai Lama has said, "high levels of compassion are nothing but an advanced state of that self-interest."

Davidson has zeroed in on a corollary to happiness—"generating compassion toward those with whom we may have had conflicts and those who are less fortunate." Because the brain of a neuroscientist resided in the body of this Occidental meditator, Davidson began to think about "how we can approach the construct of compassion and bring it into the scientific lexicon."

Easier said than done. Compassionate behavior does not immediately lend itself to rigorous quantification. How would you define it, much less measure it? Ricard, for example, told me that Buddhists practice no less than three distinct forms of compassion meditation: "Unconditional Love and Compassion," "Altruistic Love with Object," and "Compassion with Object." In terms of meaningful experiments, how do you precisely differentiate these states? And if they are truly distinct, do they also produce distinct signatures in brain activity? Is it even worth the bother to try to obtain a metric for such a fuzzy entity?

From the psychological point of view, the clear, though challenging, answer is yes. Compassion has been an essential component in contemporary psychological conceptions of wisdom, especially in the Clayton-Birren and Ardelt wings of wisdom research. It is inextricably bound to philosophical notions of leading a good life—not just a life good for oneself, but also a life virtuous in its larger social interactions. So, despite the inherent definitional difficulties (and skepticism of colleagues), Davidson has devoted a good deal of time over the last decade to probing the biology of compassion, and he has managed to publish his findings—on long-term, expert meditators as well as novices—in top-tier scientific journals like *PNAS* and *PloS Biology*. In its most recent incarnation, this exploration has focused on the world's most preeminent practitioners of compassion.

"These are the Olympic athletes, the gold medalists, of meditation," Davidson told me a few years ago, explaining his interest in these exemplary meditative minds. As he and colleagues wrote, in a 2004 *PNAS* paper as notable for the touchy-feely language it introduced to serious neuroscience as for the rather rigorous data it reported, the research attempts nothing less than to peer into the brain and reveal the neural machinery underlying behaviors like "unconditional loving-kindness and compassion"—behaviors as central to the missions of Jesus and the Buddha as they are to our more modern psychological conceptions of wisdom.

Davidson sees this work as a logical extension of other areas of contemporary neuroscience. "Just as we have done for other complicated mental processes in cognition, such as attention or memory or imaging," he told me in his office, "we're doing the same thing, trying to decompose these broader constructs into more elementary components. For a quality like compassion, there are different components. One component is adopting the perspective of another in some sense. A second component is at least feeling *something* in the 'heart,' and I use that as a metaphor for a somatic, or bodily, representation of another individual's suffering. It's an emotional response to suffering, which serves as a motivator or catalyst, if you will, propelling a person to do what he or she can to relieve suffering. A third component has to do with propelling into action. It's one thing to experience something like the suffering of another or feel it emotionally. But to actually act on it is a differentiable component, we believe." (If you recall the model of decision making proposed by neuroscientists like Paul Glimcher, with its two-step process of valuation and then choice, you can see compassion as a powerful emotional vote during the process of evaluation that precedes an actual decision).

Davidson and his colleagues believe that each subcomponent of compassion has a separate neural compartment, and over the past few years, they have begun to sketch out a preliminary map of the location of these distinct modules in the brain. "The perspective taking," Davidson explained, "is represented in an area called the temporal-parietal junction." (This is a "border" region that runs more or less horizontally between the temporal lobe, which, as its name implies, lies just beneath the temples, and the parietal lobe, which covers the rear top of the head; a quick and dirty way to think about this geography is that when you scratch behind your ears, you're at the latitude and longitude of the temporal lobe, and when you scratch the top of your head, you're in the neighborhood of the parietal lobe.) "The emotional response," Davidson continued, "is primarily centered on the insula." (This, too, is a "border" region, deep in a fissure separating the frontal and temporal lobes.) "And the translation-into-action component is represented in areas of the brain that we know to be involved in the integration of motivation and action, particularly the basal ganglia. So those are three components of a circuit that we find to be activated in these long-term practictioners when they generate compassion."

Now, linger on that word *practitioner* for a moment. It implies practice, of course, and this simple but powerful idea holds profound implications for wisdom—and for anyone who aspires to have it. A growing number of studies, Davidson's chief among them, suggest that the human mind can be *trained*—in this instance, trained in certain forms of meditation—to enhance one's powers of compassion and other mental qualities, including emotional regulation. Jon Kabat-Zinn, like Ricard trained as a molecular biologist and now a physician specializing in stress reduction at the University of Massachusetts Medical Center, has pioneered an ambitious program of mindfulness meditation training; he, too, has published peer-reviewed studies suggesting that meditation not only can improve emotional well-being but, in a collaboration with Davidson, can even enhance the performance of the immune system. Coming from a different, more clinical direction, psychotherapists like Zindel V. Segal of the University of Toronto have demonstrated that cognitive "practice"—in the guise of the repetitive cognitive exercises that are central to cognitive-behavioral therapy—clearly alters a person's psychological "performance."

Since we know the cultivation of wisdom itself is a difficult and complicated process, however, this is a scientific adventure likely to be cloaked in uncertainty and perhaps even vexed by spiritual mystery. During a meeting some years ago with Western scientists, the Dalai Lama noted that in order for something to be cultivated by training, according to Buddhist psychology, the "seed" must be there.

"Where does that seed come from?" one of the scientists asked.

"A mysterious source," the Dalai Lama said, laughing. "According to Buddhist tradition, the continuum of consciousness of an individual is without beginning."

Science keeps records of beginnings, of course, and in the life sciences, that record is called evolution. There have been growing hints in recent years that compassion—and its fellow traveler, empathy—may have its source in a unique constellation of cells and structures that have evolved in the central nervous system of both nonhuman primates and humans. These cells and structures are located in areas in the brain that have been tentatively associated with such mental capacities as recognizing feelings in others (especially feelings of pain and suffering), understanding the thoughts and intentions of others, and motivation. Scientists stumbled upon some of this biology, mostly accidentally, in the brains of research animals, where millions of neurons are arrayed in the primate brain, each programmed like a vast array of radio-telescope dishes, to detect similarities in actions between one animal and another.

In the early 1990s, Vittorio Gallese and his colleagues at the University of Parma, in Italy, first reported the existence of what he calls "a very strange class of neurons" sprinkled throughout a monkey's brain (and, it later turned out, in ours). In an oft-told tale (with several different tellings), Gallese was in the lab one day, killing time between experiments with a monkey whose brain had been wired with sensitive electrodes. At one point, Gallese reached for an object (he can't remember what—some accounts suggest a peanut, others ice cream); the monkey, while sitting motionless, observed this gesture, and Gallese heard the unmistakable crackle of brain cells firing, amplified through a computerized detector.

The real surprise was the location of these miked-up cells. They were located in a part of the brain known as the premotor cortex, which is not usually associated with watching but *is* associated with doing. These neurons exert motor control when monkeys reach for an object. As Gallese recently described it, "These neurons did not only fire when the monkey grasped or manipulated an object, but had the remarkable property of responding also when some other individual, be it another human being or another monkey in front of the monkey we were recording from, was performing the same action. . . . So there is a part of the brain which reacts in a very similar way when someone is doing something or watching someone else doing the same thing." Like cognitive motion detectors, these microscopic monitors, dressed up in the bright and shiny coat of cultural metaphor, are now widely known by their catchy moniker: "mirror neurons."

Think about the implications for our social awareness and behavior. It is as if, by selectively activating the neurons we would use to initiate and guide physical actions, we can "sense" another creature's intentions, "feel" another creature's physical experience. Researchers have learned that when animals (humans included) observe a partially obscured action, they nonetheless can infer what is going on, and the mirror neurons that reflect that *hidden* action become activated. When we hear a sound associated with a particular action (the crack of a peanut shell as it is broken, for example), the motor neurons involved in cracking a peanut shell light up. If we see a foot doing something, the motor neurons connected to those specific foot movements fire; if we see the mouth chewing or agape, the motor neurons that control the mouth are aroused in imitative sympathy.

Neuroscientists have found mirror neurons marbled throughout the brain—not just in the ventral premotor cortex but also in the left hemisphere, in the posterior parietal lobe, in the visual area, in the superior temporal sulcus region. "All these areas have been shown to be activated in this double fashion," Gallese explains, "when we act or when we see someone else acting." Marco Iacoboni of UCLA, who collaborated with Gallese and Giacomo Rizzolatti on many of the pioneering mirror neuron studies, has said, "Quite simply, I believe this work will force us to rethink radically the deepest aspects of our social relations and our very selves."

What do mirror neurons have to do with compassion and wisdom? That word *embody* hints at what is truly revolutionary about this avenue of research. Our perception of, and indeed our emotional reaction to, the outside world registers not only in the visual part of the brain, as you would expect, but in the *motor* part of the brain, which by neural design and engineering maintains connections between the motor cortex and parts of the body itself: the arms, the legs, the gut, the mouth, the eyes, the toes, any and every part of our anatomy that is muscled and can be moved at the brain's behest. Hence, as we study the expression of another human face, which can range from contortions of delight to a rictus of fear, we literally "feel for someone else" with the motor part of our brains, and, by extension, in the muscled parts of our body necessary for those actions and movements.

This notion of "embodiedness"—that our emotional perception of and response to the outside world are not merely cognitive and brain-based but also movement-oriented and body-based—has become hotly pursued by a number of leading neuroscientists. Raymond Dolan of University College London, in fact, sometimes begins his talks with a spirited recitation of a scene from James Joyce's *Dubliners,* when Joyce's protagonist in "The Dead," Gabriel Conroy, climbs the stairs of a hotel behind his wife, Gretta, in anticipation of sexual union when they reach their room. In Joyce's words (and Dolan's rendering), the wall between Conroy's emotions and bodily sensations essentially disintegrates, so that one merges with the other. Dolan's lab has taken this insight into the realm of empathy. In a 2004 experiment, people who experienced pain—which produced activity in the anterior cingulate cortex and insular cortex—showed activation in those same parts of the brain when they viewed a loved one subjected to the same pain. "Such awareness," Dolan has written, "is at the core of the psychological attribute of empathy, and, arguably, this attribute underpins a human disposition to altruism and compassion."

When you smell an unpleasant odor, your brain registers activity in a part that seems to track feelings of disgust. When Gallese scanned the brains of human subjects while they watched actors pretending to be either pleased or disgusted, the same area lit up when people observed disgust in other people (the specific brain area was the anterior insula), similar to Dolan's pain experiments. "So this neuron fired when I feel pain, but also when I see someone else feeling pain," Gallese points out. "Our body resonates along with the body of other individuals."

"Resonance"—this idea that our neurons bristle harmonically with an external reality—is not simply a beautiful concept that offers a possible route to explaining the biology of empathy; it likely represents an aspect of biology that has a very long evolutionary history. Charles Darwin spent hundreds of pages in *The Expression of the Emotions in Man and Animals* categorizing not only the range of emotion, but presenting an exquisitely observed catalog of the facial muscles involved in the expression of these emotions—including, surprisingly, meditation. If we're inclined to smile when someone else smiles, or frown when they frown, or contort our mouths to say "Yuck!" when someone else does it, we are basically employing our own muscles to inhabit another person's emotional state. Darwin intuitively, perhaps even wisely, understood the connection of muscles to emotion.

Why would natural selection, which ruthlessly culls out weak and passive behaviors, smile upon such wimpy, sob-sister behaviors as empathy and compassion? What happened to the selfish gene? This neural convergence of observation and action, of watching and then doing, jump-starts one of the most powerful forms of learning: mimicry. The notion that imitation is cross-wired in the brain—that the sensory system is, at the level of neurons, superimposed upon, if not merged with, the motor system—suggests that in at least one neural system, we are hard-wired for the "other-centeredness" that is so essential to wisdom. It also offers a splendid mind-reading form of social knowledge that, if cultivated, could be indispensable to wisdom.

Imitation and mimicry, moreover, play a central role in the theory of human evolution proposed by Robert Boyd of UCLA and Peter J. Richerson of the University of California at Davis, who argue in their book *Not by Genes Alone* that natural selection acts on human culture as well as on genes. In their view, imitation is an early evolutionary step in the accumulation of positive cultural adaptations—division of labor between men and women during Paleolithic hunting, for example, or the adoption of technological innovation—that can rapidly spread through a population, enhancing the *group's* ability to survive. They even suggest that the development of "other-centeredness," through imitation, was crucial to the evolutionary success of our species. "Perhaps in our lineage," they write, "the complexities of managing food sharing, the sexual division of labor, or some similar social problem favored the evolution of a sophisticated ability to take the perspective of others. Such a capacity might incidentally make imitation possible, launching the evolution of the most elementary form of complex cultural traditions."

Tania Singer, a neuroscientist at the University of Zurich, has been leading an ambitious effort to tease apart the neural circuitry of empathy, which she distinguishes from compassion. She has been conducting experiments, for example, that attempt to understand whether empathy is an automatic, almost unconscious reaction (as the mirror neuron story would suggest) or something requiring a little more cognitive editing, a brief way station of thought, during which we have to decide whether the information we're perceiving has enough emotional merit to warrant empathy. Before traveling to Wisconsin, our old friend Matthieu Ricard spent twenty-two hours over three days in an MRI machine in Maastricht, the Netherlands, while Singer's team probed his uniquely empathetic brain.

Gallese believes the degree to which our bodies' sensory apparatus "resonates" with another person's is nothing less than a quotient of our individual empathy. If you not only *know* what it feels like to be caressed or slapped in the face but actually *feel* those feelings to some degree, neurally and subconsciously, when you see someone else caressed or slapped, he says, that "seems to suggest that if the system, the embodied simulation process, the mirroring mechanism, works up to a certain level, we are able to empathize. If it works below a certain level, we become autistic to understanding how it feels to be in that particular state."

Many neuroscientists shy from gazing too fondly into the mirror system (even a venturesome sort like Richie Davidson cautioned against too hasty an embrace of the science). However the story ultimately plays out, the evidence to date increasingly suggests that our brains have the physiological goods to pay emotional, sensory attention to the actions, experiences, and experienced *feelings* of others, and in some cases to experience those feelings as if they were our own. That is a system endowed with the marvelous hardware to use compassion to guide judgment, and if wisdom in part arises from a similar attentiveness to, and feeling for, others, then perhaps we have an innate capacity for compassion. Not necessarily an innate *abundance* of compassion, mind you, but, rather, the neural hardware with which to cultivate and attain compassion and, perhaps, a bit of wisdom.

And cultivation of compassion, finally, may be the ultimate frontier of this research. In his latest round of experiments, Davidson has recruited laypeople to undertake a two-week, online, self-taught program of compassion meditation training. Study participants were asked to imagine the suffering of another individual and then to think about relieving that suffering (covering all bets in the office pool, the experiment asked participants to practice compassion for "a loved one," a friend, a stranger, and even "a difficult person"). A mere fortnight of compassion training produced, surprisingly, discernible changes in brain activity, particularly in the structure known as the insula, which has cropped up again and again in this first wave of social neuroscience experiments as an important emotional barometer, tracking empathy, injustice, compassion, and other aspects of "social weather." Other prominent scientists are working the compassion/meditation circuitry: Antonio Damasio's group recently attempted to distinguish between the neural correlates of admiration and compassion, and James Gross's group published a study showing how "even just a few minutes" of loving-kindness meditation increased what they called "social connectedness."

As we've stated before, these are very much early days for the neurobiology of compassion and empathy. But Davidson's focused, reductive research on meditation conceals a much bigger ambition, which he hinted at in a question he posed recently at an academic presentation: "Are there strategies that may influence a person's affective style and transform the neural circuitry of emotion, and of emotional regulation, into a more positive direction?" Davidson would probably be loath to state this idea more baldly, but I am happy to provide one reasonable translation. Are there ways to train our minds to be more compassionate and, by extension, wiser?

If that sounds impossibly grandiose, especially in translation, consider the perspective of Matthieu Ricard. In his hermitage outside Katmandu, where his perch for meditation faces an awesome panorama of Himalayan peaks, he has spent tens of thousands of hours training himself to be compassionate and mentally focused. As we sat one afternoon in the second-floor atrium of the neuroscience building at the University of Wisconsin, facing the rather more prosaic panorama of a parking lot and a soccer field, he spoke for quite a while about wisdom. Ricard pointed out that meditation is a form of cognitive exertion every bit as demanding as weight lifting or long-distance running, and he expressed surprise that more people don't give their minds a proper workout. "It seems odd," he said, "that we have this incredible brain and yet spend so little time cultivating it."

CHAPTER EIGHT

HUMILITY

The Gift of Perspective

The more the mango tree flourishes, the more it droops.

—Indian proverb

If someone tells you that somebody else is saying awful things about you, don't defend yourself against the accusations, but reply, "He must not know about the other faults that I have, if these are the only ones he mentioned."

—Epictetus

IN MARCH 2009, against a backdrop of cultural outcry and newfound economic austerity, a Manhattan auction house offered for purchase some of the last surviving personal effects of Mohandas K. Gandhi. The items included a banged-up pocket watch, a simple bowl and plate, the famous steel-rimmed spectacles, and a worn pair of sandals—just about everything left from his threadbare wardrobe but the trademark loincloth. Those powerful props continue to remind us of Gandhi's ascetic lifestyle. As *The New York Times* put it, "They were the simple belongings of a man who did not care for possessions." But there is a wonderful anecdote in Louis Fischer's biography of Gandhi that suggests it wasn't always so.

In the fall of 1888, when he was only eighteen years old, Gandhi sailed to England to commence law studies. Born in 1869, the future prophet of nonviolent protest came from relatively humble origins (*gandhi* means "grocer" in Gujarati) and his early life hardly augured the greatness he would later achieve. Gandhi, fragile in both mind and body, recalled running home from school every day "lest anyone should poke fun at me." He married at age thirteen, later dropped out of college in India, and became an absentee father with his departure for England, all the while conspicuously handicapped by a debilitating shyness. Gandhi nonetheless determined to become a lawyer, passed the entrance exams, and gained admission to law school in London. Shortly after he began his studies, however, an English friend expressed concern about the failure of his Indian colleague—thin, always dressed in inappropriate clothes, and a vegetarian in roast beef–eating England—to fit in. Despite what he called "the all too impossible task of becoming an English gentleman," Gandhi agreed to undergo a makeover. He signed up for dancing lessons, studied elocution, and bought a violin, the better to appreciate Western music. And he went shopping for clothes on Bond Street. Hence we get this priceless word picture of Gandhi standing in Piccadilly Circus about a year later later, on a day in February 1890.

Gandhi, according to his friend Sachchidananda Sinha, "was wearing at the time a high silk top hat 'burnished bright,' a stiff and starched collar, a rather flashy tie displaying all the colors of the rainbow, under which there was a fine striped silk shirt. He wore as his outer clothes a morning coat, a double-breasted waistcoat and dark striped trousers to match, and not only patent-leather shoes but spats over them. He also carried leather gloves and a silver-mounted stick but wore no spectacles. His clothes were regarded as the very acme of fashion for young men about town at that time. . . ." It's not clear what happened to the silver-tipped cane, but it was certainly not among the items recently put up for auction.

Gandhi's splendid *Project Runway* turn in Piccadilly seems at odds with the enduring picture we have of the man in the loincloth, yet it also captures the paradox of humility, which, in turn, is embedded within the paradox of wisdom. This short-lived makeover was motivated not by vanity, but, rather, by a desire to repay the kindness and concern of Gandhi's English friend with an attempt to "become polished and [cultivate] other accomplishments which fitted one for polite society." The experiment lasted about three months. "If my character made a gentleman out of me, so much the better," Gandhi concluded. "Otherwise I should forego the ambition." By trying on a personality that didn't truly fit, he rediscovered the person he truly was; his brief career as a clotheshorse reintroduced him to his core values. And it reminds us that humility, like wisdom, often begins with self-awareness, especially the awareness of one's own limitations.

The theme of humility recurs again and again in Gandhi's prolific writings. As a child, his intense social awkwardness (which might well be diagnosed today as pathological shyness), his frail constitution, and his self-perceived cowardliness all contributed to an unusual reticence, which Gandhi managed to convert into a strength: "My hesitancy in speech, which was once an annoyance, is now a pleasure. Its great benefit has been that it has taught me the economy of words. . . . Proneness to exaggerate, to suppress or modify the truth, wittingly or unwittingly, is a natural weakness of man and silence is necessary in order to surmount it." Most of all, Gandhi used his innate humility as a guide for subsuming himself—his ego, his sensual needs, his material desires—in a much larger sense of mission.

But here is the paradox: Gandhi's promulgation and practice of non-violence as a mass political movement—first as a young (well-dressed) lawyer in South Africa, later in the great twentieth-century fight for Indian independence—seems anything but humble. It required the heart of a lion in its dedicated challenge to authority, a passionate belief in truth, the resolve not to back down, even when confronted with physical intimidation and abuse. The Gandhian paradigm captures the inner strength of humility, which makes it such an unusual and often invisible component of wisdom. He used humility as a ferocious focusing device, stripping away every pleasure that might distract—he called this the "annihilation of one's self," a powerful echo of the Buddha—from what was important. Humility allowed him to marry his deep moral sense to action, and to convince himself (and others) that the power of his conviction could not be deterred. So even though Gandhi himself attributed his humility to reticence, frailty, and struggle, we can recognize it as well as a source of incredible strength.

The link between humility and wisdom is so deep, so enshrined in literature and philosophy, that it's a wonder that science has taken an active interest only in its evil twins, narcissism and pathological shyness. There is a reason that this is the shortest and indeed thinnest chapter in a section devoted to science. Humility has attracted relatively little empirical research interest (although that is changing), and for a reason that even Gandhi himself intuited. "Truth and the like perhaps admit of measurement," he wrote, "but not humility. Inborn humility can never remain hidden, and yet the possessor is unaware of its existence."

Humility as a virtue essential to wisdom has deep roots. One of the earliest (and most amusing) passages is Plutarch's account of the golden tripod. This was ancient Greece's equivalent of an Oscar for being the wisest person in the world during the seventh century B.C., and none of the nominees would be caught dead accepting it. It was initially bestowed upon Thales of Miletus, but he declined to accept and passed the hot potato along to Bias, whom Thales deemed to be wiser. Bias, too, demurred. Eventually, the honor made the rounds to all Seven Sages of the ancient world, and all seven refused to accept it. They recognized that to acknowledge the honor would instantly, in Gandhi's phrase, annihilate it, like some philosophic form of antimatter. So from the very beginning of human agency, wisdom and humility have been yoked.

Citing the etymological origins of a word has become a dreary cliché, but as a gardener, I can't resist mentioning the roots of *humility*. The word derives from the Latin *humilis,* which is usually translated simply as "humble." But I'm drawn to the metaphorical richness of one of its alternative and less common meanings, "humus," which bespeaks a rich, organic, nose-to-the-ground, bottom-up, of-the-earth pragmatism that aptly captures its origins, its sense of scale, and its point of view. This "groundedness" is reflected in a Tibetan saying: "Humility is like a vessel placed at ground level, ready to receive the rain of qualities." Despite getting a little etymological dirt under our collective fingernails, it's also important to view humility in somewhat more abstract terms. Kant celebrated it, according to the scholar Jeanine Greenberg, as "the moral agent's proper perspective on himself as a dependent and corrupt but capable and dignified rational agent." Humility, to Kant, was one of the essential virtues.

Modern philosophers seem less sure. In *The Stanford Encyclopedia of Philosophy,* for example, the philosopher Sharon Ryan considers, and then rejects, what she describes as the "Humility Theory" of wisdom. She recounts the story of Socrates' serial exposures of inflated self-regard in Athens in the fifth century B.C. and then boils down its message into the kind of logical construct that philosophers often use: "S is wise iff S believes s/he is not wise." Ryan rejects this idea for several reasons, among them the entirely reasonable argument that "many people who believe they are not wise are correct in their self-assessment." Ryan ultimately concludes, "Humility views of wisdom are not promising." In fact, she suggests exactly the opposite: "We should hope that a wise person would have epistemic self-confidence, appreciate that she is wise, and share what she knows with the rest of us who could benefit from her wisdom. Thus, the belief that one is not wise is not necessary for wisdom." But this misses the point so forcefully made by the way Gandhi "shared" his wisdom: His campaign for truth (and change) gained moral authority, and thus social traction, in part because of his obvious humility.

And this is as good a place as any to draw an important distinction between religious and secular concepts of humility, because this distinction spills over into larger cultural notions of wisdom. Wisdom and humility are deeply entwined in the biblical literature: Moses was "very humble, more than any person on the face of the earth," according to a passage in Numbers (one of the earliest examples in literature equating humility with wise leadership), and in Saint Peter's first epistle, he exhorted Christian believers to be "compassionate, lovers of the brethren, merciful, humble. . . ." But the dominant religious message attached to humility, from the Wisdom literature of the Bible through the writings of theologians in the Middle Ages, is about piety and obedience to God. By the thirteenth century, Saint Thomas Aquinas argued that humility "consists in keeping oneself within one's own bounds, not reaching out to things above one, but submitting to one's superior."

This theological version of humility is part of religion's repeal of *human* wisdom during medieval times; in placing inordinate emphasis on obedience, fealty to God, and self-denial, it inevitably shaped behavior and made a virtue of submission. This is hardly just a Christian tic. The theme of humble obedience and submission crops up repeatedly in Confucius, too; he harps on proper, discreet behavior with superiors and ascribes lesser qualities to common people (or, as he calls them, "Small Men"). And yet was there a more serenely disobedient spirit in all of religious history than Jesus? And though he was deeply religious, Gandhi's entire political philosophy grew out of civil disobedience. How do we reconcile the submissive humility promoted by religion with the self-assured humility associated with wisdom?

If we consider obedience in a secular or, even more narrowly, behavioral sense, it may help explain why humility persists as a virtue. It is one of those traits that acts as a social lubricant, greasing the wheels of group interaction, minimizing interpersonal friction, enhancing the odds for cooperation. And in order for one's social group to function smoothly—whether that group is a family, a community, a congregation, or, yes, a mass movement for national independence or civil rights—cooperation is essential. The working parts of humility are self-deprecation of a non-pathological sort, humor, an antipathy to greed, and a cosmic perspective. Humility may be the ultimate social emollient. The paradox, again articulated by Gandhi, is how we come by humility. "Truth can be cultivated as well as Love," he wrote in a letter. "But to cultivate humility is tantamount to cultivating hypocrisy." In other words, you have to be, not *try* to be.

So this poses a conundrum. As Sharon Ryan reminds us, we want our sages to be confident, self-certain, and, in Horace's ringing invocation, daring. Poor Richard frequently extolled the virtue of modesty, but Benjamin Franklin, writing in his *Autobiography,* pointed out, "Most people dislike vanity in others, whatever share they have of it themselves; but I give it fair quarter wherever I meet with it, being persuaded that it is often productive of good to the possessor, and to others that are within his sphere of action; and therefore, in many cases, it would not be quite absurd if a man were to thank God for his vanity among the other comforts of life." At the same time, it's no accident that the Seven Sages viewed the golden tripod as an albatross. Humility is a quality that demands careful titration, a delicate (and often shifting) balance between personal agency, social deference, the inner strength of self-awareness, and a good-humored grace in acknowledging human limitations.

There is a wonderful anecdote in Plutarch's life of Pericles, as wise and revered a leader as ancient Greece ever produced, that captures this delicate balance (indeed, tension) between accomplishment and humbleness. One day, as he attended to some urgent business in the marketplace, Pericles was assailed by a heckler. This "idle hooligan" followed and tormented him all day long without cease. "Towards evening he returned home unperturbed, while the man followed close behind, still heaping every kind of insult upon him," Plutarch wrote. "When Pericles was about to go indoors, as it was now dark, he ordered one of his servants to take a torch and escort the man all the way to his own home."

It is impossible to imagine such a thing happening in contemporary society for many reasons, central among them the suspicion that virtually all our public servants are much too prideful and vain to put up with even a nanosecond of spontaneous challenge, much less an entire day's worth of comeuppance. Some people might regard this episode as humiliating to a great leader like Pericles; it strikes me, to the contrary, as a measure of his greatness. Humility here, as in the example of Gandhi, required uncommon emotional strength, intellectual conviction, and serene self-confidence. As the French philosopher François de La Rochefoucauld observed, "Few people are wise enough to prefer useful criticism to treacherous praise."

As Gandhi pointed out, humility defies measurement. But in the last ten years, a handful of researchers has begun the daring project of attempting to do so. Once again, it is the psychologists and the behavioral economists who have led the way.

In 2007, two economists at Pennsylvania State University's Smeal College of Business published a study that, indirectly, says a lot about humility in the executive suite. They had set out to explore the relationship between the performance of companies in the computer business and the narcissism of their respective chief executive officers. In order to wrap their arms around this admittedly elusive quality, Donald C. Hambrick and Arijit Chatterjee cataloged several indirect measures of corporate vanity in order to gauge the dimensions of any given CEO's ego: the size of the leader's photograph in company documents, the length of entries in *Who's Who,* the frequency with which the CEO was mentioned in corporate press releases, and the number of times the CEO used the first-person singular (I, me, mine, my, myself) in interviews.

This is obviously a squishy metric, but reading their analysis (entitled "It's All About Me") through the prism of the 2008 financial panic and the 2009 executive bonus boondoggle, we may find the results eerily informative. Narcissistic CEOs, they reported, were more acquisitive, tended to make more "extreme choices," and, in essence, were the corporate equivalent of drama queens, typically pursuing abrupt shifts in business strategy. Hambrick and Chatterjee wrote that "CEO narcissism is also related to extreme and volatile company performance. Narcissistic CEOs, who tend to pursue dynamic and grandiose strategies, also tend to generate more extreme performance—more big wins and big losses—than their less narcissistic counterparts (as measured both by accounting and shareholder returns)." The bottom-line conclusion: Executive narcissists bring a lot of drama into the boardroom, but that drama does not produce systematically better corporate performance.

Narcissistic CEOs may, however, contribute to financial panics. In an interview, Hambrick dated the rise of boardroom narcissism to, of all things, a *publishing* event in the 1980s. "It all got launched by Lee Iacocca and his autobiography," Hambrick said, referring to the iconic executive of the (now bankrupt) Chrysler Corporation. "Prior to that, ninety-five percent of the American public would not have been able to name the CEO of *any* company. So there has been a spiral of flamboyance, of more narcissistic types drawn to these jobs, to the kind of compensation that would attract that kind of person. Everything became more extreme, jazzier, more colorful. And this," he added, speaking about a month after the collapse of the U.S. financial industry, "is where it's left us."

Perhaps the best explicit treatment of humility in the executive suite (and in almost all the social sciences) has come from management researcher Jim Collins. In his 2001 book *Good to Great,* and in an article in the *Harvard Business Review* based on the same research, Collins's findings in a sense ratified the same paradox apparent in Gandhi. Exceptional business leaders, according to the five-year study, blend "extreme personal humility with intense professional will." Business executives who "possess this paradoxical combination of traits are catalysts for the statistically rare event of transforming a good company into a great one." Among the traits associated with these humble (but great) business leaders were a reluctance to seek public adulation; a style of leadership that emphasized organizational standards rather than personal charisma; a willingness to subsume personal ambition to the larger ambitions of the company; and an eagerness to apportion credit to others combined with a willingness to accept responsibility for failures. Collins anticipated the Penn State findings when he noted that "boards of directors frequently operate under the false belief that a larger-than-life, egocentric leader is required to make a company great." It's worth commenting on the methodology here, too: Collins's research project began by identifying eleven companies that significantly outperformed the general market and also competitors in their field; it was only in an attempt to explain why these companies had performed so exceptionally that the pattern of humble but strong-willed executive leadership emerged.

Like aggression (but not compassion), narcissism (but not humility) has always been a legitimate object of academic study—first identified as a psychological characteristic by Havelock Ellis in the nineteenth century, later legitimized by Freud, and by now part of the small change of popular culture. But beyond the white noise of psychobabble, psychologists have at least made some progress in their attempts to measure narcissism. Since 1979, with the development of the Narcissistic Personality Inventory (NPI), psychologists have used a questionnaire methodology to measure narcissism as an aspect of personality. A considerable amount of psychological research over the last thirty years has produced a picture of the narcissist as a person with inflated self-regard, feelings of superiority and entitlement, and a ravenous, almost insatiable hunger for attention and praise—values clearly antithetical to almost anyone's notion of wisdom.

"We know that narcissism correlates with lower ethics in business and with white-collar crime," said W. Keith Campbell, a psychologist at the University of Georgia who studies narcissism. "What it really does predict is the emergence of leadership, in groups and among strangers. But over time, it's humility, coupled with drive, that correlates with *performance.*" He added that narcissism has a pervasive effect on decision making "that's generally bad." As Chatterjee and Hambrick put it, "Narcissists rate themselves highly (and more highly than is objectively warranted) on an array of agentic dimensions, including intelligence, creativity, competence, and leadership abilities."

Many of us don't need the NPI to reach the same conclusion; we seem to walk around with invisible, imaginary, yet remarkably precise social calipers that tell us when someone has a swelled head. It is much harder to measure an unswollen head, but that's not to say social scientists aren't trying.

Julie Juola Exline, a psychologist at Case Western Reserve University, is among a handful of academics who have been attempting to lay a ruler against the idea of humility. Having cut her sociological teeth working in the field of narcissism (Exline has collaborated with Campbell), she became interested in the positive aspects of what is sometimes known as a "quiet ego." In 2004, with colleague Anne L. Geyer, Exline took a first tentative step—not unlike Vivian Clayton's initial work on wisdom—by conducting a questionnaire survey of college students, asking them their attitudes toward humility. "Somewhat surprisingly," they reported, students held "quite positive views of humility," seeing it as a psychological strength and associating it with a pro-social modesty about achievements and abilities. Several other research teams are now trying to develop a reliable scale for measuring it.

Despite the dearth of hard data, the theoretical underpinnings of these fledgling humility studies nonetheless bear importantly on the subject of wisdom. June Price Tangney, who contributed a chapter on humility to the *Handbook of Positive Psychology,* has sketched out the key features of humility, which resonate with many *psychological* traits associated with wisdom: an ability to acknowledge limitations and mistakes, an openness to new ideas and new contradictory knowledge, a knack for avoiding self-aggrandizement, an ability to keep one's achievements in perspective, and the kind of self-aware self-perception that perceives both strengths and weaknesses. As Exline noted in a recent paper, "A humble self-view need not be negative or self-deprecating, and it does not require a sense of inferiority to others. A humble person might see the self as a relatively small part in a larger scheme of things, perhaps in comparison to God or to the universe." Gandhi made the same point more poetically when he wrote, "Has an ocean drop an individuality of its own as apart from the ocean?"

These are obviously preliminary and unsatisfying observations, and the road ahead is nettled with potential potholes. Humility research has located itself in the field of positive psychology, which remains controversial and seems to provoke considerable skepticism in hard-core scientists. There is also the evolving definition of humility as viewed through a religious lens. The late Sir John Templeton, whose foundation has funded some of the early research on humility, wrote with passion about it as a virtue and distanced himself from earlier, more abject versions of its meaning. "Humility is not self-deprecation," he said. "To believe that you have no worth, or were created somehow flawed or incompetent, can be foolish. Humility represents wisdom. It is knowing you were created with special talents and abilities to share with the world; but it can also be an understanding that you are one of many souls created by God, and each has an important role to play in life." He also pointed out, "Humility leads to prayers as well as progress and brings you in tune with the infinite."

Although it is as difficult a subject as wisdom in its own way, humility is a potentially rich area for empirical inquiry. It may speak with a paradoxically loud voice about our behavior in groups, with implications for cooperation, altruism, and all the social spheres of activity in which wisdom plays such an important role. "To the extent that it helps people to transcend self-interest," Exline has written, "humility should also pave the way for virtues such as forgiveness, repentance, and compassionate love."

"Love is the strongest force the world possesses," Gandhi wrote, "and yet it is the humblest imaginable." In 1921, out of a sense of responsibility to the thousands of Indians who discarded, at his behest, their expensive foreign-made garments and piled them upon pyres of protest, the one-time Piccadilly dandy announced his intention to wear only a loincloth and chador (or shawl). "I adopt the change," he explained, "because I have always hesitated to advise anything I may not myself be prepared to follow." This, too, is submission, but to a common goal that was anything but submissive.

Humility may leave less of an empirical paper trail than any other component of wisdom because it is the one most in thrall to mystery. By thrall, I don't mean submission so much as appreciation. In an odd way, humility forms, like a crystal, in the "mother liquor" of limitation. To be more precise, it is the only residue of temperament that is possible after you've expended the tremendous Socratic energy necessary, using all the things that you know, in order to discover all the more that you don't know. We *graduate* to a humbled understanding that so much information—about the nature of people and the nature of their interactions, about the foundation of decisions and the prediction of future actions and events—remains so inaccessible and therefore so profoundly unknowable that it is only fitting to respond with humility in the face of such immense uncertainty. (Even though, as anyone who's read the *Apology* must concede, Socrates did an awfully poor job of concealing his own arrogance, even as he exposed everyone else's limitations.)

Viewed this way, humility straddles two foundational aspects of wisdom: the limits of knowledge and the acknowledgment that change and uncertainty is a natural state of affairs. You can practically hear Montaigne chuckling as he wrote, "So vain and frivolous a thing is human prudence; and athwart all our plans, counsels, and precautions, Fortune still maintains her grasp on the results."

To some modern psychologists, the humility that comes with limitation is the ultimate distillation and definition of wisdom. In an extended 1990 essay on this theme, John A. Meacham argued that "the essence of wisdom is to hold the attitude that knowledge is fallible and to strive for a balance between knowing and doubting." Like many others, Meacham draws a sharp distinction between mere factual knowledge and true wisdom. "To be wise," he wrote, "is not to know particular facts but to know without excessive confidence or excessive cautiousness. Wisdom is thus not a belief, a value, a set of facts, a corpus of knowledge or information in some specialized area, or a set of special abilities or skills. Wisdom is an attitude taken by persons toward the beliefs, values, knowledge, information, abilities, and skills that are held, a tendency to doubt that these are necessarily true or valid and to doubt that they are an exhaustive set of those things that could be known." Humility is a central component of that attitude.

Another contemporary psychologist, Robert J. Sternberg, converges on the qualities of humility by invoking another word: *sagacity,* or exceptional discernment. In a study published nearly twenty-five years ago, Sternberg noted that sagacity refers to a suite of behaviors we typically associate with humility: considering advice, learning from other people, admitting mistakes, reflecting often, being a good listener, and acknowledging multiple perspectives on an issue. Note, these qualities do not define humility; rather, they are behaviors often associated with a humble personality. They are part of the connective tissue of behavioral traits that increase the probability of wisdom.

Like wisdom itself, genuine humility is seemingly so universal that we recognize it, both instantly and gratefully, whenever we see it—especially in the public arena, where its absence is so conspicuous. That is where the humility of Gandhi, and the enormous spiritual strength it subserves, silently reprimands all of us for the triviality of individual self-importance. "Interdependence is and ought to be as much the ideal of man as self-sufficiency," he wrote in 1929. "Man is a social being. Without inter-relation with society he cannot realize his oneness with the universe or suppress his egotism. . . . Dependence on society teaches him the lesson of humanity."

As I was writing this, I was happily reminded of the power of humility by one of my son's class projects. Each February at the school he attends, all the fourth graders, girls and boys alike, don beards, bow ties, and stovepipe hats and then circulate throughout the school as "Tiny Lincolns." They memorize arcane details of Abraham Lincoln's difficult childhood, recall his commonsense cunning as a country lawyer, and, of course, recite as much of the Gettysburg Address as the memories of impatient nine-year-old minds will permit. As a result, we heard a lot of recitation of this speech in the kitchen, and although everyone remembers "Four score and seven years ago" in the beginning, fewer people probably remember the one-sentence grace note toward the end, where Lincoln said, with majestic economy, "The world will little note, nor long remember what we say here, but it can never forget what they did here."

It isn't that Lincoln humbly downplayed the significance of one of the greatest pieces of political oratory in history; there is nothing strategic about this remark, other than its artful redirection of civic attention, and gratitude, to where it belongs. The deceptively simple syntax of this complex sentence is other-centered, flowing from the egoless, nonroyal "we" (there's no I, me, my, mine, or myself in this CEO's vocabulary) to "they": "what they did here." The directional arrow of Lincoln's regard is outward; the humility is genuine. He sincerely believed that his words, or any words, were nothing compared to the ultimate sacrifice others had made. And if you accept Gary Wills's argument that those 272 words "altered" the Constitution and "undertook a new founding of the nation," Lincoln lubricated the pill of intellectual revolution with the rhetoric of humility.

As my son also reminded us, Lincoln spoke for only three minutes, following the featured orator, who spoke for the better part of two hours. The genius of true humility, and its arterial attachment to wisdom, is in understanding the context of the moment, understanding one's audience, and, at the same time, misunderstanding (perhaps deliberately) one's own greatness. The following day, back in Washington, Lincoln took a moment to scribble a note to Edward Everett, the long-winded speaker who had preceded him in Gettysburg, praising his "eminently satisfactory" discourse and adding, "I am pleased to know that, in your judgment, the little I did say was not entirely a failure." Lincoln signed the note, as usual, "Your obedient servant."

CHAPTER NINE

ALTRUISM

Social Justice, Fairness, and the Wisdom of Punishment

> *Do not do to others that which angers you when they do it to you.*"
>
> —Isocrates, "To Nicocles"

> *Nations are sustained neither by wealth nor by armies, but by righteousness alone. It is the duty of man to bear this truth in mind and practice altruism, which is the highest form of morality.*
>
> —Mohandas Gandhi

ALTRUISM IS A DISTINCTLY SOCIAL (and biologically paradoxical) aspect of wisdom. It is, as Gandhi suggests, personal morality writ large across a society, a selfless and at times self-sacrificing devotion to a greater good, whether the social unit is a crowd, a hive, a community, an institution, or a nation. It heeds the higher call of social justice, which is why Abraham Lincoln and Nelson Mandela and Martin Luther King, Jr., and Gandhi are so often elevated by popular acclamation to the conclave of the wise. Now scientists are suggesting that altruism has a pair of microscopic loudspeakers planted on a ridge high in the brain, like stereo chips empowering that Gandhian sense of doing the right social thing.

Before we get to neuroscience, I want to begin this journey in the wilderness, the one traversed by King David in his eventual journey to Jerusalem. When I first began to consider the role of altruism in wisdom, I followed the same path as Vivian Clayton, who made a literary pilgrimage to many of the ancient tales of wise behavior immortalized in

the Hebrew Bible and ancient Greek literature. As Clayton quickly realized, the Old Testament literature teems with prudent kings and just
judges, often celebrating the decisions and actions of these "wise men"
(they were usually men). The actual *process* of these judgments—both a
deep, dispassionate appreciation of what was right and an equally deep,
other-centered compassion that meted out punishment judiciously—
not only has a great deal to do with traditional notions of altruism and
social justice but, as I began to learn, intersects with some very recent scientific thinking about the central *neurological* importance of altruism.

To revisit an earlier theme, consider the "wisdom of Solomon" from
a social justice perspective. In its popular distillations, the story of
Solomonic wisdom almost always boils down to the anecdote of two
mothers arguing over a disputed infant; the familiar tale is often viewed
as antiquity's supreme example of an overarching, enlightened form of
social wisdom. But the "wisdom trajectory" of Solomon's life is actually
much more complicated, as was hinted at earlier, and tells us that the
achievement of social justice requires a selflessness that few mortals can
sustain over the course of a lifetime.

How wise was Solomon? Upon the death of King David in about
965 B.C. (roughly half a millennium, therefore, before the era of Socrates
and Confucius), his crown passed to his son Solomon—but not without
the dysfunctional family politics and political intrigues that infuse so
many biblical tales. Adonijah, another of David's sons and a scheming
rival, claims title to the throne and is celebrating his rise to power with
cronies when word spreads that the king has instead anointed Solomon.
Trembling in fear of retribution and cowering for protection near an
altar, Adonijah pleads with his brother for his life. And Solomon
responds in an evenhanded and compassionate fashion: "If he proves to
be a worthy man, not one of his hairs shall fall to the ground; but if
wickedness is found in him, he shall die."

Here we have the makings of a challenging decision! Should Solomon
allow his older brother, a presumptuous and potentially dangerous political rival, to live? Or should he, for the benefit of his nascent tribe but
also for himself, slay his rival? Does he possess the wisdom to discern
between his own self-interest and that of the nation he has now been
entrusted to lead? And does he possess the wisdom to discern between
worthiness and wickedness in a person against whom he clearly has a

bias? You can imagine the competing neural values flooding into *his* decision maker.

As it turned out, his scheming brother made the decision easy for Solomon. Adonijah advanced a seemingly simple request: If he couldn't have the throne, he said, he should at least be allowed to take his father's concubine as his wife. Now when King David was near the end of his life and "could not get warm," his servants had summoned a beautiful young virgin named Abishag to lie in his bosom and bring him warmth. Some biblical scholars read Adonijah's interest in Abishag not as the pathetic last lustful grab of a royal wannabe, but, rather, as a clever, back-door legal trick to grab power, relying on ancient precedent suggesting that marriage to a king's concubine opened a potential line to royal succession.

Solomon apparently interpreted it that way, too. He saw in Adonijah's request the seed of wickedness he'd been looking for, and he had him slain. In fact, he had many of Adonijah's allies killed to consolidate his power. Solomon's little wave of terror smacks more of bloodlust and hunger for power than wisdom, but you could at least argue that by eliminating his enemies, the young king quelled a potentially divisive succession battle, and nominally improved the odds of survival for his fledgling group of Israelites. Indeed, following the purge, Israel embarked on an unprecedented era of peace and prosperity. By punishing the would-be rule benders, Solomon achieved great good for his people.

That is usually not what people think of when they think of Solomonic wisdom, of course. His transformation from a shrewd, if vengeful, young king to a ruler with "very great wisdom, discernment, and breadth of understanding as vast as the sand on the seashore" occurred only after he was visited by God in a dream. "Ask what I should give you," God said in the dream, and after much humble hemming and hawing, Solomon replied, "Give your servant therefore an understanding mind to govern your people, able to discern between good and evil." (He asked, in short, for exactly the same power of discernment that the serpent offered Eve in the Garden of Eden.) And God said, "Indeed I give you a wise and discerning mind; no one like you has been before you and no one like you shall arise after you."

Only then, shortly after awaking from the dream, was Solomon confronted by the two prostitutes, each claiming to be the mother of a tiny infant. By threatening to cut the child in half to satisfy their demands, the king was able to discern the real mother by her anguished cries, and thus was able to "execute justice." In the book of life that is biblical literature, wisdom and social justice form their own double helix, passed down from generation to generation in self-replicating moral tales.

Even though Solomon's wisdom is divine and therefore ultimately derivative, it provides a kind of behavioral blueprint for clarity in social decision making. It requires understanding (which we might think of as the acquisition of pertinent information), discernment (using that knowledge to make critical distinctions and evaluations), and action (making a judgment based on that evaluation), all in the interest of achieving justice. Everything is guided by an underlying sense of what is right and fair. But the real surprise here is that so many of Solomon's judgments, including the ones before the dream, executed justice through punishment—a word we don't customarily associate with wisdom. And if we think of the sword as a metaphor for punishment, we can see (as biblical scholars have) that part of Solomon's life-span development was the progression from *use* of the sword to the *threat* of its punitive use.

The paradoxical coupling of wisdom and punishment may surprise us a little at first, but it is not surprising at all to a small but influential group of academics—economists, anthropologists, behavioral psychologists, and neuroscientists—who have been pondering the punitive aspect of altruistic behavior over the last fifteen years. They believe that punishment is essential to the smooth functioning of a social group and, by extension, a society. In fact, they argue that social punishment is an intrinsic feature of altruism.

Ernst Fehr, a Swiss economist and behavioral scientist at the Institute for Empirical Research in Economics at the University of Zurich, believes this innate discernment of what is fair, and how to enforce social justice within a group—the exercise of communal wisdom, if you will—must have evolved very early in the prehistory of the human race. In fact, he believes it is so central to social cooperation that it emerged earlier than language, and may have been even more important than language in shaping the earliest human expression of social wisdom.

"What is more fundamental, the evolution of language or the evolution of cooperation?" he asked when we met for coffee one morning in New York. "In my view, the evolution of cooperation is *the* fundamental thing, because you can have cooperation before you have language." In fact, naturalists have known for a long time that other species, which do not possess language, nonetheless learned the virtue of altruism.

Charles Darwin was terrified of bees. Not in the normal way—he wasn't physically afraid of being stung. Rather, he confessed deep unease about how certain perplexing aspects of bee behavior threatened to sabotage the foundations of natural selection and would thereby topple the theory of evolution.

Altruism thrives, Darwin knew, in that teeming insectopolis known as the hive. There are sterile castes of individuals, the worker bees, who willingly sacrifice their lives to protect the queen, and have no possibility of reproduction. This odd, self-negating form of altruism, Darwin acknowledged, posed a "special difficulty, which at first appeared to me to be insuperable, and actually fatal to the whole theory." If genes were, in Richard Dawkins's famous formulation, "selfish," how to explain the persistence of a fatally unselfish, altruistic behavior (and, like all behaviors, with a genetic component to it) in social insects like bees, wasps, and ants? Why would natural selection embrace behavior that by definition represented an evolutionary dead end?

The answer, Darwin began to grasp (but only partially), probably had to do with what he called "family selection"—that is, individual traits that conferred survival advantage to the family unit as a whole, if not to every individual within it. It took about a century to fill in the gaps, biologically and mathematically, but by the 1960s and 1970s, evolutionary biologists led by William D. Hamilton and Robert Trivers began to recast human altruism—the tendency to help others, at cost to oneself—in biological terms. Hamilton came up with an equation for "kin selection" that explained how the absence of genetic self-interest could still be "Darwinian"; in 1971, Trivers introduced the notion of "reciprocal altruism," an idea that reduced social relations between organisms (including humans) over time into a long-running saga of tit-for-tat interactions that allowed behavioral room for both cooperation and punishment. The biology of altruism has inspired scientific popularizers (Dawkins's books and Matt Ridley's *The Origins of Virtue* are particularly elegant examples), and a prominent British economist-philosopher recently wrote in *Nature* that the evolutionary explanation of altruism "will surely come to be seen as one of the major breakthroughs of twentieth-century science."

But a humanist with his nose pressed against the window of science might reasonably ask, What took so long for biologists to get it? Back in the day—a day in China, sometime around the turn of the sixth century B.C.—a young man named Tzu-kung asked Confucius, "Is there any single saying that one can act upon all day and every day?" The Master replied, "Perhaps the saying about consideration: Never do to others what you would not like them to do to you."

Altruism lies at the heart of Confucian notions of goodness. Indeed, it lies at its semantic heart. D. C. Lau, a Chinese scholar who has translated the writings of Mencius, a disciple of Confucius, makes the point that *gen*—the central concept, the mother word of the Confucian Way—has multiple meanings in English: not just goodness, as it is usually translated, but also benevolence, human-heartedness, love, humanity, and, yes, altruism. Confucius's ancient version of the Golden Rule finds obvious echoes in Buddhism, in the teachings of the prophet Muhammad, in the philosophy of Greeks dating back to Thales of Miletus, and, of course, in the Judeo-Christian tradition, crystallized in Jesus's admonition "Do to others as you would have them do to you."

These are ancient observations, but who could have imagined that they would have surprisingly modern biological implications? The syntax of every variation of the Golden Rule is a humanistic, nonmathematical formula for social reciprocity. The version of the Greek philosopher Isocrates (it serves as one of the epigraphs to this chapter) is particularly modern, because it acknowledges the role of anger, explicitly connecting our innate sense of fairness with an emotional state. Confucius's notions of social judgment allow the "gentleman" (his role model for wise conduct) to use both reciprocal *and* sanctioning behavior. "Where gentlemen set their hearts upon moral force, the commoners set theirs upon the soil," Confucius said. "Where gentlemen think only of punishments, the commoners think only of exemptions." In this black-and-white Confucian world, there are rule makers and rule breakers.

Even economists understood the importance of altruism, fairness, and social punishment before biologists. Exactly one hundred years before Darwin published his theory of natural selection, the youthful Adam Smith wrote in his first book, *The Theory of Moral Sentiments* (1759), "Nature has implanted in the human breast, that consciousness of ill-desert, those terrors of merited punishment which attend upon its violation, as the great safe-guards of the association of mankind, to protect the weak, to curb the violent, and to chastise the guilty."

"Terrors of merited punishment" sounds like a secular, Industrial Revolution knockoff of biblical wrath, but Smith located this fear in the human breast, about twelve inches south of its true domain in the human brain. Geographical coordinates notwithstanding, Smith's insight went largely ignored, even by economists, until about twenty years ago. That is when a Swiss academic began to investigate the social repercussions of not paying a working Joe a decent wage, thus embarking on an avenue of research on fairness that ultimately led back to biology and right into the brain.

"Throughout most of its history," Ernst Fehr acknowledged recently, "mainstream economics has relied on the simplifying assumption that material self-interest is the *sole* motivation of *all* people, and terms such as 'other-regarding preferences' were simply not part of economists' vocabulary." And, it's probably fair to add, throughout most of *its* history, mainstream biology didn't care what economists thought about material self-interest (or altruism).

But beginning in the early 1990s, some experimental economists like Fehr began to step back and take a broader view of human decision making. They designed experiments that probed not only economic self-interest but also economic selflessness, or altruism. Those experiments have pushed altruistic decision making deep into the brain, and have begun to identify neural circuitry that, to some degree, comes into play when we attempt to discern fairness or inequity in a social situation. As anyone who has ever held the short straw knows only too well, this sense of unfairness plugs right into our emotions.

The neuroscience of altruism grew in large part out of an obscure issue of economic inequity known as "involuntary unemployment." In the late 1980s, Fehr toiled in relative anonymity as a labor-market theorist, exploring an esoteric but long-standing aspect of fairness in the workplace. His question can be boiled down to this: Do employees work hard if they are paid what they perceive to be an unfair wage? Common sense—and anyone who has worked for the minimum wage—would suggest that this question has significant economic and social implications. But, as Fehr discovered, mainstream economics (with the notable exception of John Maynard Keynes, decades earlier) had largely ignored the issue of fairness. "Just like other people," he wryly noted in a recent essay, "most economists appreciate being treated fairly in their personal interactions—and would be deeply unhappy if paid less than what they consider to be fair. But when it comes to the modelling of economic affairs, fairness is not a currency that counts." Fehr worked up a paper, arguing that employers strive to keep employee wages at the going rate for a very simple reason: They know workers paid an unfair wage are not as productive. Despite this seemingly mild observation, every economic journal to which he submitted the paper rejected it.

Fehr, however, had stumbled onto something much bigger than paycheck politics, and, as he put it, "I searched for ways to capture the issue of fairness empirically." Like other experimental economists, he began to use game theory and the latest technology, including fMRI and other state-of-the-art detectors of brain function, to probe the way humans wrestle with issues of social justice and equity. Before long, his research about social fairness was appearing not just in the *Quarterly Journal of Economics* but in *Nature* and *Science,* as well.

As Fehr and researchers of a similar bent began to discover in the 1990s, a sense of fairness is central to many of the decisions we make, including (but not limited to) economic behavior. Fairness concerns, he noted, influence a broad array of human interactions, including business relationships between companies, trade negotiations, the allocation of public goods, and the politics of resource management, not to mention a "decisive" role in the evolution of human cooperation. In other words, Fehr's initial curiosity about inequitable wages exploded into a wide-ranging inquiry into the nature of human altruism.

Altruism, in its simplest rendering, is the antithesis of selfishness. It can be a form of selflessness in an interaction with another person, or it can reflect behavior that promotes the welfare of an entire group. Scientists concede that there are, in fact, many complicated motivations for altruistic behavior, and not all of them are flattering; they range from selflessness to heightened compassion to a desire to burnish one's reputation.

But there also seems to be an innate human desire to cooperate. In experiments that use game theory to explore human altruism, researchers have shown again and again that people, on average, initially tend to cooperate—at least until they are betrayed by their opponent. Once betrayed, most people then revert to a more guarded strategy of cooperation—a tit-for-tat form of reward and reprisal—which rewards another person's cooperation with cooperation and punishes selfishness with selfishness. This picture of tit-for-tat cooperation, which initially emerged in computer simulations of game theory, led to the concept known as "strong reciprocity." Popularized by Samuel Bowles and Herbert Gintis of the Santa Fe Institute, it refers to the deeply ingrained human habit of rewarding helpers and punishing cheaters.

Whether you're an economist or a biologist, an obvious question—perhaps *the* obvious Darwinian question—about altruism is, What's in it for the person who sacrifices advantage or incurs personal cost in order to benefit another or others? One answer is that altruistic behavior enhances one's reputation in the community, just as selfish behavior is likely to diminish the willingness of others to cooperate with you. This is especially true of the ongoing relationships that occur in a complex social group—a continuum of interaction that game-theory experimenters attempt to model as serial rounds of the same game. But burnishing one's reputation does not merely have a social payoff; you may hit the Darwinian jackpot of reproductive success, because the ultimate evolutionary advantage attached to altruistic behavior may be reproductive attractiveness. Unlike those in any other animal society, humans may use altruistic behavior to "signal," as biologists put it, somewhat invisible but highly attractive traits that make them more desirable mating partners, and more desirable allies when conflict is in the offing.

But reputations can take a long time to establish and are always in flux—van Gogh became a genius long after he died, and both he and Jesus struck most contemporaries as more than a bit mad. What's in it for the *individual* who acts out of altruism? One answer began to emerge in 2002, when neuroscientists scanned the brains of people engaged in a game that tests one's commitment to social cooperation. So let's play Prisoner's Dilemma.

In this classic game, two participants ("suspects") sit in separate "cells" and are offered several choices by their jailers. If one cooperates with authorities and betrays the other, the betrayer goes free and the other person receives a ten-year sentence; if both remain silent (which turns out to be a form of cooperation between the players), both must remain in jail, but only for six months; and if both try to betray each other, they both must remain in jail for five years. So for anyone in a long-term relationship (anything other than a one-off encounter), the smart (dare we say wise?) behavior is mutual cooperation. But there's still a price to pay, so what's so great about cooperation?

In a study published in 2002, James K. Rilling and his colleagues showed for the first time that people who exhibit mutual cooperation (that is, remain silent and refuse to betray the other player) during the game, and see their own cooperation rewarded by the other player, produce a distinct blush of activity in the reward centers of the brain—the same part of the brain, in other words, that lights up during reinforcement learning. The activity is driven by the neurotransmitter dopamine, tickling receptors deep in the mid-brain (specifically in the striatum) and in the orbitofrontal cortex. Put crudely, this is the brain's way of saying that social cooperation provides the same kind of neural kick as sex, drugs, and rock 'n' roll. Who knew?

What makes the dopamine connection to altruism truly fascinating is how intrinsically social, how "up close and personal," cooperation needs to be to set off fireworks in the brain's reward circuitry. When experimental subjects play Prisoner's Dilemma with a computer, for example, the brain's reward circuitry basically yawns with disinterest when it encounters digital versions of cooperation; when you interact with a cooperative but inert liquid crystal display, the thrill is gone. Altruism seems to require the flesh-and-blood presence of another person to trigger the reward centers (and thus the heightened neural attention that accompanies cooperation).

These initial experiments on altruism led to a bigger issue that forms part of altruism, cooperation, and, I would argue, wisdom: punishment. Solomon's early behavior, before God visited him in the dream and granted him wisdom, may, in fact, have been wiser than we originally thought. Ruthless and punitive though it appears, his decision to eliminate his rival Adonijah—whose attempt to circumvent the rules of royal succession can be seen as a biblical version of "free riding"—allowed the nascent nation of David to flourish. Perhaps this extreme form of punishment came at a cost to his reputation, but by eradicating his rivals, Solomon paved the way for setting up a cooperative, highly organized, and ultimately successful nation.

So King Solomon's ruthlessness may tell us something important, in broad brush, about fairness and social justice. In order for altruism to work, it requires an element of constraint (if not outright punishment) of cheaters and free riders. It appears to have deep evolutionary roots (researchers have identified altruistic behavior in monkeys). And we can see a combination of altruism and punishment in a bunch of theoretical games refereed by tweedy academics. But does any of this correspond to real events in everyday life? In my recreational reading (I'm a football fan), I ran across an anecdote that convinced me that altruistic punishment may, in fact, be responsible for the phenomenal success of one of the world's best-known sports enterprises.

In 2006, the National Football League reached agreement with five separate television networks on a five-year, $24 billion broadcasting contract; the record deal generated, each year, a combined $3.7 billion for the league, with the proceeds distributed equally among thirty-two franchises. That's a far cry from the first contract in 1962, a deal with a single network that brought in $4.6 million. But when that first contract was negotiated in 1961, the terms by which the money was shared were determined by a real-life version of what economists call the "Ultimatum Game"—a game that Ernst Fehr and other economists have frequently used in experiments to study altruistic punishment.

The Ultimatum Game works like this: One player, the "proposer," is given a sum of money (say one hundred dollars) and is allowed to decide how much she or he wants to share with a second player, the "receiver." The catch is that if you are the receiver and you feel the proposer's offer is insultingly low, you can reject it as unfair, in which case neither of you receives anything.

Humans have a predictable threshold for perceiving unfairness in this exchange; many studies have shown that when the proposer's offer drops to around 25 percent of the total (twenty-five dollars in this example) or less, the receiver typically rejects it as unfair, and everyone ends up empty-handed. The receiver not only denies himself some money (thus making it altruistic) but denies the proposer his unfairly larger share as well (thus making it altruistic *punishment*). These results contradicted classic economic theory, because to deny oneself a reward (even a small one) in order to deny an even bigger reward to another has traditionally been seen, at least by economists, as irrational behavior. Indeed, it is game theory's version of cutting off your nose to spite your face.

A remarkably similar scenario unfolded during the NFL's league meetings in Miami in 1960, when the owner of a big-market team, Dan Reeves of the Los Angeles Rams, got into an argument with a young upstart executive from a small-market team, Dan Rooney of the Pittsburgh Steelers, over how television revenues would be divvied up. "If there's going to be a package deal," Reeves insisted, according to Rooney's recent autobiography, "then the big markets should get more money. There's a couple of ways we can do this."

"There's only one way we can do this," Rooney shot back, "and that's to divide the money equally."

"You'll never get the votes to pass that—it's just not going to happen!" Reeves thundered.

Rooney, however, argued that a package deal, equally shared, would generate more revenue for the league (a prediction that has proven true) and would allow small-market teams to remain competitive on the field and solvent off it (a prediction also borne out). But the big-market teams insisted on their right to a bigger share. The two men argued back and forth over the revenue issue, and at one point, the Pittsburgh owners threatened to cut off broadcasts to Los Angeles when the Rams played in Pittsburgh. "Then you won't get any money," Reeves pointed out.

"Then neither will you!" Rooney responded. "You won't get a dime, and you've got more to lose."

Functional MRI machines had not even been invented when this conversation took place, but if scientists had taken a peek at Dan Rooney's brain at this juncture, they probably would have seen a pattern similar to that of the hundreds of experimental subjects who have been consumed by feelings of unfairness while playing the Ultimatum Game. Indeed, in 2003, Alan Sanfey and his colleagues provided the first glimpse of a well-defined neural circuitry that comes into play when our sense of injustice is provoked by the Ultimatum Game. They saw heightened activity in the dorsal striatum (where rewards are registered) and in the insular cortex, two small knots of cortical tissue (one on each side of the brain) that seem to be involved in the perception of injustice and disgust; activity also perks up in the anterior cingulate cortex, which is a kind of neural barometer of internal conflict. (Like almost any brain activity, it should be added, this pattern of arousal can be increased or decreased, depending on emotional state, memory of prior events, and the like.)

When you look at the bigger picture—a picture that includes the well-being of a social group over an extended period of time—Rooney's behavior doesn't seem irrational at all. In fact, it seems exceptionally . . . wise. By almost universal consensus, the agreement to negotiate a single contract and share the revenues equally has helped the NFL to become the most successful sports enterprise in the United States, if not the world. It has promoted competitive vigor, economic health, and, perhaps most important to the league's reputation (and therefore its brand), the perception by fans that the on-field competition is fair. So Rooney's "punishment" of Dan Reeves—or, to be more psychologically accurate, his *threatened* use of the economic sword—in their real-life version of the Ultimatum Game actually conferred significant advantage to his "group," the NFL.

And that, ultimately, is the whole point of cooperation, which is really another name for social wisdom. As Harvard University biologist Martin A. Nowak puts it, "From hunter-gatherer societies to nation states, cooperation is the decisive organizing principle of human society." Most important, Nowak says cooperation increases the "average fitness" of a population, which is another way of saying that natural selection smiles on groups that get along. And it all begins with a sense of fairness.

When I described the NFL episode to Ernst Fehr, he seemed delighted to learn that altruistic punishment appeared to play such a positive role in American sports culture. "There is an interdependency here that makes it reasonable to argue that the deal can't be proportionate, where a big city like L.A. gets more and a small city like Pittsburgh gets less," he said. "There's a true sense of fairness involved."

The Rooney anecdote represents a real-life example of how exercising a wise form of altruism holds real-life benefits for a community in the long run. Sacrificing one's short-term gains for the long-term benefits of the group (the group could be a family, a community, a business, a sports league, or a nation) is arguably among the wisest of human behaviors. But altruism calls upon a diverse suite of cognitive and emotional skills: discerning the fundamental unfairness of a situation; understanding the value of cooperation; having the courage to defy one's own immediate self-interest, with the aim of achieving a larger goal; and having the patience to wait for the rewards of that larger goal to materialize. All those skills reflect neural processes in the brain, and although some will be discussed in other chapters, the notion of altruistic punishment has emerged—particularly in the research of Fehr, Bowles, Gintis, and Joseph Henrich of Emory University—as critical to cooperation in human society.

In order for a society to function cooperatively, be it a hunter-gatherer tribe or a nation-state, its members or leaders must sanction those who break the rules. Fairness is the crucial litmus test. In prehistoric times, the cheaters were probably those who helped themselves to a share of food, while not contributing their fair share of effort to the hunt (here, too, the situation is a little more complex: In times of plenty, cheating probably doesn't deprive the group in a deep material sense, but, rather, may create small but unacceptable tears in the social fabric that widen into group-threatening holes during times of want). In modern times, and especially in the experimental setting of the Ultimatum Game and other artificial behavioral challenges, cheaters tend to be monetarily selfish, taking more of the pie, sharing less of the endowment.

How closely do these games reflect day-to-day social interactions? As Fehr and his colleague Urs Fischbacher noted in a 2003 paper, "A key element of the enforcement of many social norms, such as food-sharing norms in hunter-gatherer societies, is that people punish norm violators not for what they did to the punisher but for what they did to others." In other words, in real social interactions, cheaters tend to be punished not for their crime against an individual, but, rather, for their transgression against the interests of the group at large.

Do biology and wisdom converge on this point? Formally, no, because the experiments (at least now) would be too hard. But Robert Boyd, professor of anthropology at UCLA and a deep thinker on the evolutionary role of culture, was willing to connect some important anecdotal dots in an interview. "We have lived in cooperative societies, where sociopaths do poorly, for a long time," he said. "Why aren't we like baboons? If the dominant male in baboon society beats you up, that's it. But we have this culturally transmitted moral system that includes a lot of things that people think of as wisdom."

Is it any wonder, then, that the foundational texts of ancient wisdom—those of Confucianism, Buddhism, Islam, Christianity, Judaism, Greek philosophy, you name it—devote an extravagant and eye-glazing amount of space to defining, often in mind-numbingly meticulous and quasi-legal fashion, the norms of proper behavior? An enormous amount of philosophic and theological energy has gone into understanding human nature well enough to know how to spot cheaters and where to look for clues to aberrant antisocial behavior.

This area of research has recently produced one other big biological surprise: When people engage in altruistic punishment, the same part of the brain that becomes aroused by cooperation (the reward center of the dorsal striatum) turns on as we're being punitive. We get a neural kick from both cooperation *and* punishment. In a fascinating experiment in 2006, Fehr and his Swiss colleagues showed that it is even possible to knock out, temporarily, the reciprocal part of people's intrinsic sense of "reciprocal fairness" (our social justice "module," if you will). How? By zapping a specific part of the brain while people play the Ultimatum Game.

As part of this experiment, participants were deliberately tendered insultingly low offers. As the subjects stewed in a self-marinating sense of injustice, researchers aimed a precise (and, by all accounts, harmless) pulse of electromagnetic energy known as transcranial magnetic stimulation (TMS) at the part of the brain that normally calls for retaliation when the game reaches this point (the right dorsolateral prefrontal cortex). Remarkably, the participants *knew* that the offers were fundamentally unfair, below the threshold of acceptability. Nonetheless, they were *neurologically incapable* of acting altruistically on that sense of unfairness; they failed to reject the unfair offers because of the external TMS interference in a cognitive part of the brain.

Fehr's group has helped pioneer the use of these brain-jamming techniques to tease apart the neurobiology of altruistic punishment, and the experiment established that, in effect, an emotional part of our brains (the insula) discerns unfairness (at least the unfairness of this reductively simple game), while a more cognitive part of the brain (the PFC) wheels into action to redress the social injustice. One way of looking at this, as Fehr and his colleagues do, is that the cognitive part of the brain is necessary to override selfishness, economic self-interest, and, if we want to be blunt about it, greed.

As any parent knows, part of wisdom is knowing when to dispense punishment, how much to give, when to let things slide. How do we calibrate the right dose? And does this "wise" behavior at the level of family interactions in any way correspond with the grand philosophic preoccupation with social equity and distributive justice that is such a central component of our larger collective social wisdom?

Two recent experiments suggest the immense social and philosophical implications of the neuroscience of altruism. In 2008, researchers at the University of Illinois and the California Institute of Technology reported the results of a clever experiment about social inequity and distributive justice that attempted, among other things, to peel back the neural curtains on two long-standing debates in the history of philosophy. One involves the tension between the greater good for the many (or utilitarianism, as articulated by John Stuart Mill), and the supreme rights of each person (or deontology, as the philosopher John Rawls argued). The other debate concerns whether our sense of justice is rooted in reason, as argued so exhaustively by Kant and Rawls, or in the kind of moral sentimentalism that the young Adam Smith proposed more than two centuries ago (readers will recall from the chapter on moral judgment that this is the same difference as in the debate between reason and emotion in moral decisions).

The researchers attacked these age-old questions by using a scenario involving the allocation of relief money to groups of children living in an orphanage in Uganda. They scanned the brains of twenty-six subjects (the average age was about forty) as they struggled to decide whether it was better to give a little aid to a lot of children or, conversely, to give more meaningful aid to a smaller group of children. This dilemma pitted equity (every child's "right" to help) against efficiency (the most meaningful and "helpful" form of help, even if it didn't reach everyone), which is the core tension of our desire to spread justice equitably and well.

Indeed, it is the kind of practical social (and often political) decision that is Solomonic in its demand for wisdom.

The MRI scans revealed a division of neurological labor as the subjects grappled with this kind of problem. The perception of social inequity heightened activity in the insular cortex; that was no surprise, as it had been suggested in earlier altruism experiments. However, a brain struggling with meaningful distribution of justice also showed activity in a small region known as the putamen, which appeared to weigh the efficiency of different actions. Both regions sent along their relative evaluations of the allocation relief plan to a third area (for the record, the caudate/septa subgenual region) for a "unified measure of efficiency and inequity."

What should particularly interest us about this experiment, however, is not the circuitry per se, but the idiosyncrasy within those circuits from person to person. Different individuals showed significantly different sensitivities to injustice and efficiency; where those differences originate is unclear, but there is a lot of personal variation in the neural processing of social injustice, so much so that there is no "right" way to approach these vexing dilemmas. "More broadly," as the authors (Ming Hsu, Cedric Anen, and Steven R. Quartz) wrote in *Science,* "our results support the Kantian and Rawlsian intuition that justice is rooted in a sense of fairness; yet contrary to Kant and Rawls, such a sense is not the product of applying a rational deontological principle but rather results from emotional processing, providing suggestive evidence for moral sentimentalism." In other words, the early, sentimental Adam Smith appears to have been right about the roots of fairness.

This enterprising research, this dialogue between traditional philosophy and new brain science, is pushing us toward an understanding not only of individual behavior but of what might be considered the wisdom of crowds and even the wisdom of institutions. In 2006, researchers in Europe published the results of an experiment showing that even a selfish, greedy mob ultimately sees the wisdom of a social system that exercises altruistic punishment. The experiment is a little complicated, but bear with me, because the denouement is very surprising.

The researchers—at the University of Erfurt in Germany and the London School of Economics—recruited dozens of participants and asked them to choose membership in one of two "institutions"; as part of their membership, each participant would receive an "endowment," a certain amount of money. One of the institutions was advertised as observing sanctions (both rewards and punishments), and the other operated without any sanctions. Once the participants had chosen their institution, the experiment entered a second phase: People in both institutions were asked to play a "public goods" game, a classic dilemma pitting individual self-interest against group welfare. In this case, participants could voluntarily contribute a portion of their endowment to the public good in the sanction-free institution, but they faced punishment in the other institution if they didn't contribute.

As the game unfolded, nearly half the people in the sanction-free institution turned out to be free riders, and contributions to the public good eventually dropped to zero. Although the sanction-free institute boasted higher material benefits for its members at first (since the greedy held on to their money), these advantages quickly tailed off. In the more punitive institution, meanwhile, payoffs started out being lower, but the punishment of deadbeats eventually meant higher payoffs to everyone who had contributed. Over time, people in the punitive institution did much better.

In the third and most astounding aspect of the experiment, participants were asked, after each of the multiple rounds of the game, if they wanted to switch membership from one institution to the other, having observed how the two institutions functioned and evolved during the earlier phases of the experiment; *virtually very single member of the sanction-free institute, including the cheaters, chose to migrate to the institution where strong reciprocity reigned.* This was the game-theory version of the criminal who wants to be caught; even the cheaters recognized wisdom when they saw it in a punitive but ultimately successful institution, and "fully depopulated" the institution without norms.

The good news: There is wisdom in this collective decision to choose a system rooted in altruistic cooperation (and punishment). The bad news: Greed and free riding essentially ruined one society (sound familiar?) until collective wisdom kicked in. The stakes could hardly be larger in the non-make-believe world in which we live.

Anthropologist Joseph Henrich has pointed out that "the puzzle of cooperation in large groups" bears on everything from dealing with global climate change and valor in combat to voting in elections and donating blood. "Such cooperative dilemmas, or 'public goods' problems, involve situations in which individuals incur a cost to create a benefit for the group. . . . The dilemma arises from free-riders who enjoy the group benefits created by the contributions of others without paying the costs." And, as he points out, social cooperation can collapse if the cheaters proliferate too much.

Social cooperation is indeed fragile. When researchers run these "public goods" experiments as a one-time exercise, they find that participants are typically cooperative, contributing between 40 percent and 60 percent of their "endowment" to the public good. But when scientists run a variation of the game without the possibility of punishment, they discover that cooperation quickly disintegrates. Altruists are initially optimistic about the contributions of others, but when it becomes clear that there are a lot of free riders and cheaters in a group, the altruists, in effect, become disillusioned and pessimistic, minimizing their contribution to the public (or common) good, resulting in what Fehr and others call a "decay in cooperation." Fehr has shown theoretically that a small number of selfish individuals can sabotage a population of predominantly "strong reciprocators."

Altruistic punishment, however, completely rewrites this greedy script. A minority of tit-for-tat operators, if given a chance to punish social defectors and noncooperators, can deter an increase in free riders, even though the cheaters may represent a majority in the population at first. This "corrective" aspect of altruistic punishment, moreover, finds surprising resonance with modern concepts of wisdom. A fundamental quality of wisdom, according to philosopher John Kekes, is that it is corrective, and since "wisdom is corrective, it is exercised only when correction is needed, and that happens when things do not go smoothly."

The idea that we derive as much satisfaction from punishing cheaters as we do from selfless cooperating may not seem like wisdom. But as Ernst Fehr told me in a conversation, "Wisdom is always relevant to the culture of social interactions."

Finally, let's expand the frame here. Not all punishments are altruistic, of course, and we need an even higher standard of wisdom to understand that the very notion of altruistic punishment can be subverted or corrupted in its execution. One of the reasons Solon won acclaim as one of the Seven Sages of the ancient world is that he repealed the extremely punitive ("draconian") laws of his predecessor, Draco. And when he was asked which city he considered the best governed of all, he purportedly replied, "The city where those who have not been wronged show themselves just as ready to punish the offender as those who have been." That indeed is a wise community.

The followers of Confucius similarly delighted in the downfall of the coercive and punitive Ch'in dynasty in 206 B.C. In the Ch'in ethos, discipline and order were so strict that any officer who was late in performing his duties faced summary execution. So when a Ch'in officer, charged with transferring a group of convicts, realized that he was running late and would therefore be killed, he decided to release his prisoners, then led a rebellion, which ultimately toppled the dynasty. "If the emperor had inculcated loyalty by humaneness, rather than attempting to control his people with draconian punishments, the Ch'in Dynasty might not have been so short-lived," writes Confucius scholar Sarah Allen.

Most of the important and difficult decisions we have to make in daily life are not simple either/or choices weighing narrow monetary gains in simple games; indeed, they are important and difficult precisely because they are complex and ambiguous, and the rewards may not be apparent for many years to come. Many of these decisions do not, as in neuroeconomic experiments, require an answer in a matter of seconds. And these results lull us into a false geography of human behavior, where one tiny quadrant of gray matter is the "center" of fairness, another the "reward center," yet another the place where we act on the basis of judgments about social justice.

There is a danger in seeing this as a map of isolated points rather than a three-dimensional, pulsing, dynamic network of neural coordination, one that is constantly changing, and changeable, one that is weighted with different inputs depending on our previous experiences, our learning, our mood that day, the general uncertainty or anxiety we may be feeling, our life circumstances at the time, our age and stage of life— a network that is, in a word, idiosyncratic. A truly adaptable machine (and none has ever been invented to rival the adaptability of the versatile human brain) must solve the engineering problem of using a finite number of working parts to adjust to an almost infinite number of problems as if our life depended on it, which, of course, it does—and, in terms of the species, has for hundreds of thousands of years. "The brain is a very ingenious and flexible device," Fehr told me. "And our sense of fairness must have a representation in the brain."

If we are all evolutionarily endowed with the ability to discern fair from unfair, and derive deep biological satisfaction from both altruism and altruistic punishment, why don't we see more evidence of altruistic wisdom around? Why don't we leverage these basic biological impulses for fairness and for public goods into much greater and longer-lasting acts of social and political cooperation? The story of Solomon's wisdom offers a sobering valedictory on this point, too.

Despite his receiving the gift of God's wisdom, the reward center in Solomon's brain began to respond more strongly to other, less altruistic pleasures. Biblical scholars note that the king spent twice as long building his own home in Jerusalem as in constructing the temple to God; his taste for gold was so great that silver became a mere common metal in his realm. The biblical account of his house, with its four hundred carved pomegranates, its twelve bronze oxen supporting a massive bath, and its elaborate wall paneling of imported cedar, reads like *Architectural Digest* (Jerusalem edition) or *Lifestyles of the Rich and Biblical.*

One by one, Solomon violates virtually every norm of kingly behavior that God had meticulously spelled out in Deuteronomy. When he finally flouts the rule against taking too many wives, it is the final straw. God decides to punish Solomon for behaviors that anyone (except perhaps a libertarian or neoclassical economist) would recognize as self-interest run amok: He has become an acquisitive, lustful, and arrogant tyrant who flouts the rules. Solomon the wise has become Solomon the free rider; God finally lowers the boom. And what form does the punishment take?

It is not Solomon who suffers directly, but his *kinship group.* "Since this has been your mind and you have not kept my covenant and my statutes that I have commanded you," God says, "I will surely tear the kingdom from you and give it to your servants." And indeed, God bestowed all but a small piece of Solomon's kingdom on Jeroboam, his servant, with the words, "I will take the kingdom away from his son and give it to you." The hired help walk off with the vast majority of Solomon's kingdom, and his heirs are deprived of all but a small portion of his wealth.

Solomon's descent into terminal foolishness is also a story about the perishability of wisdom, the impermanence of altruism, the importance of humility in sustaining both, and, ultimately, the constant, daily, life-long, almost inhuman effort required to muster goodness (*gen*) or, as economists put it, other-regarding behavior. We tend to regard wisdom as an armchair activity, but the ancient sages recognized how much relentless effort it requires. When one of his disciples asked Confucius about the key to wise government, he replied, "Lead them; encourage them!" When the disciple pressed for more specifics, the Master added a single word to the prescription: "Untiringly."

CHAPTER TEN

PATIENCE

*Temptation, Delayed Gratification, and the Biology
of Learning to Wait for Larger Rewards*

> *The ego isn't an organ that might sit in a central
> place like Descartes's pineal-based soul. It's a net-
> work server, a broker of cooperation among the
> interests, and like interests, is itself engendered
> and shaped by differential reward—specifically,
> by the long-range reward that comes from better
> defense against short-range rewards.*
>
> —George Ainslie, *Breakdown of Will*

> *Did I not keep my nerve, and use my wits to find
> a way out for us?*
>
> —Odysseus, in Homer, *Odyssey*

IT IS ONE of the most psychologically iconic scenes in all of literature:
While a Mediterranean wind fills the sail of their boat with a "canvas-
bellying breeze," hurtling them toward the island of the Sirens, Odysseus
gathers his crew and explains the mortal danger they are about to face.
Those famously bewitching songs will lure any impulsive mariner to his
doom. Odysseus longs to hear the fantastic melodies for which the Sirens
are renowned, but has also been warned that they "will sing his mind
away on their sweet meadow lolling." The songs would be a pleasure to
hear, but, alas, a pleasure fatal to all his long-term interests and goals.

If, as some philosophers maintain, human agency (and, therefore, human wisdom) first found its voice in the epic poems of Homer, then Book 12 of the *Odyssey* presents the original case study for a lot of modern psychological and, surprisingly, microeconomic thinking about the nature of patience, impulse control, and the daily struggle inside the human mind to make the right and prudent choice about the future when immediate temptations threaten to divert us from our best destination.

The dilemma that plays out in Odysseus's head is a classic bipolar shouting match between two conflicting interests, a time-related battle that has long fascinated philosophers, psychologists, neuroscientists, and even economists. One is a short-term interest: Odysseus's desire to hear the Sirens' song (which is synonymous with an urge to wreck his ship while trying to hear more). The other is a long-term interest: to return home to his wife and children. In the long run, getting home to one's family is much more important than listening to a pretty song. But in the short run, the song seems much more alluring. In the marketplace of the brain, where these values are tested and weighed against each other at the level of molecules and algorithms, whichever choice offers the greatest value will trip the decision maker, but here's the neurological wrinkle in this ancient conflict: As you get closer to the island, the value of hearing the Sirens' song suddenly seems much larger than getting home. Nearing a short-term reward creates, in the words of one scientist, an "unexpected warp" in the way we perceive the value of things.

If wisdom is in part the decisions and actions that arise out of knowing what is most important, how do we overcome this built-in warp in perception? In Odysseus's case, a successful resolution begins with the understanding that the urge to listen to the Sirens' song, irresistible as it seems, would sabotage his larger, long-term plans. Understanding this, he has to adopt a strategy to outwit his impatient, impulsive self. This he does by stuffing the ears of his crew with sun-softened beeswax (bees again!) and instructing them to tie him to the mast of the ship; no matter what he says under the influence of the Sirens, the men are to ignore his wild gesticulations and lash him even tighter to the mast. Put simply, his imagined future self—at home at long last with his wife, Penelope, and his children—has to outsmart his present self in order to reach the future of his imagination. Metaphorically speaking, it is the same battle we fight when we resist a chocolate that will sabotage a diet, resist a drink that will ruin our sobriety, resist buying that wide-screen television in order to save for a rainy day, resist the flirtation that will ruin our marriage. To George Ainslie, who has studied this phenomenon in everything from rats and pigeons to humans and Homer, each of these dilemmas boils down to a debate between our present and future selves.

Odysseus's dilemma about the Sirens is not simply an anecdote about impatience from literature. Jon Elster, a psychologist who has done groundbreaking work on the biology of addiction and choice of reward, wrote an entire book about it (*Ulysses and the Sirens: Studies in Rationality and Irrationality*). And Ainslie, whose early experiments on pigeons at Harvard University in the 1960s laid bare the biology of this kind of choice, has returned to the example of Odysseus again and again in his writing to illustrate the kind of free-for-all that occurs inside our brains as we weigh immediate and delayed rewards, a process known as "intertemporal bargaining."

It was Ainslie who first understood that this conflict could be plotted as a graph that predicted human behavior, how that graph was anticipated by Socrates, how the internal mental conflicts predicted by that graph run through the entire literature of human frailty, from Homer through Saint Augustine and Shakespeare to every forgettable potboiler in contemporary fiction, how that graph captures every variation in the theme of self-destructive impulsivity, from violating a diet to succumbing to the urge to smoke to engaging in any of the dozens of self-defeating behaviors that each day diminish and erode our hopes for the future. The graph is called the "hyperbolic discount curve," and just as it predicts that Odysseus was destined to give in to the Sirens' temptation if he hadn't outwitted his short-term desires, it predicts both human impulsivity and the cognitive tricks that promote human patience.

Hyperbolic discounting, Ainslie has written, "sets a person against herself and makes Ulysses hearing the Sirens the enemy of Ulysses setting out for home." And because hyperbolic discounting probably reflects a fundamental neurological process, it requires a repertoire of tricks to resist temptation and exercise patience. The most important trick is what Ainslie calls "willpower." And willpower, he says, is a skill of knowing and acting that amounts to a form of wisdom.

. . .

It is a long, long way from the wine-dark sea to the Veterans Administration hospital on the outskirts of Coatesville, Pennsylvania, a grim collection of ruddy Depression-era brick buildings and clinics hunkered on a gray little slouch of a hill about fifty miles west of Philadelphia, where the exurbs dwindle into rural emptiness. Many of the veterans who come here for treatment are psychiatric patients, and one of the psychiatrists they come to see, unbeknownst to virtually all of them, is among the deepest and most global of thinkers about self-control, destructive behavior, and the human potential to make better choices.

I went to Coatesville because Ainslie's name kept cropping up in all sorts of far-flung conversations and contexts. A well-known Harvard economist referred to him as a "genius." Robert Frank, the well-traveled and wide-ranging economist and author, described Ainslie in his book *Falling Behind* as "one of the most interesting and creative people I have ever had the pleasure to know" and called his 1992 book, *Picoeconomics*, "brilliant." Leading neuroscientists with whom I spoke, including Nathaniel Daw at New York University and Peter Dayan of University College London, steered me toward his 2001 book, *Breakdown of Will*, a short, dense volume of both philosophic meditation and scientific interpretation about what motivates people to behave the way they do, which is to say badly. The book somehow manages to combine both an economic view of brain function at the microscopic level (hence "picoeconomics") and a big-picture view of human behavior that takes it into the realm of wisdom.

When I visited him at the VA hospital, George Ainslie's second-floor lair in Building 11 was more like a geologic formation than an office: Stalagmites of reprints rose from the floor, in the corners and under the table, and a sedimentary ring of shelving around the edge of the room contained, like shale, layer upon thin layer of yet more scientific papers. The bookcase at one end of the room, and the spillover into the hallway outside, betrayed all the compass points of Ainslie's sprawling scientific thought about human behavior: *Principles of Neural Science*, of course, but also works by Freud and Saint Thomas Aquinas, George Vaillant's *Adaptation to Life*, William Hazlitt's essays, and texts on behavioral economics. The eight-volume *Encyclopedia of Philosophy* lay within arm's length of the chair where he works, just to the right of his computer monitor (and right next to a framed picture of his wife).

"Self-control," Ainslie was saying, "is really the art of making the future bigger." He was talking about how Odysseus had to frame the problem of the Sirens in the marketplace of the mind, where immediate and future interests vie for supremacy. The idea of getting home has to pull on his mind, and estimating the value of that future—a value that we conjure or cognitively coddle or simply make up—is what the will does to tip the balance against impulse. Ainslie describes this process as "constructing your idea of your character, your idea of heaven, your idea of simply the moral life, the kind of person you insist on being in the long run. And that entity, that interest, exists in the marketplace, and fights off the Sirens. Wisdom isn't just an insight," he said. "It is a budgetary skill. Ulysses's wisdom is not just knowing he's better off not sailing onto the rocks; it is his knowing what to do about the chance that he *will* sail onto the rocks."

Ainslie speaks in a rapid, gravelly voice, edgy with authority, but what you most notice about him at first is his narrow bald head, fringed by a ring of gray hair, and a strikingly high forehead; he reminded me of one of those highly evolved future humans in science-fiction movies who have unusually large forebrains. Despite his view of the brain as a marketplace of competing interests, Ainslie is no friend to classical economists (he dismisses standard utility theory because he believes it fundamentally misunderstands the inherent conflict of human preferences). But behavioral economists have loved his research for decades because of its insights into human behaviors; psychologists and psychiatrists appreciate his surprising friendliness to Freudian ideas ("I always say, if Freud was a stock, now would be the time to buy"); neuroscientists have been inspired by his work to launch a new series of experiments about patience and impulsivity; and humanists should like him, because his model of human behavior allows more for the possibility of free will than other neuroscientific models. His day job, his narrow clinical interest, however, is motivation, particularly as it affects drug addiction, since any addiction is, in his view, the ultimate devaluation of the future in favor of immediate reward. These are behaviors rooted in a biology that, as Ainslie has written, shows "how people knowingly choose things they'll regret."

A series of unexpected results in animal behavior experiments led Ainslie to his economic model of willpower. Raised in Binghamton, New York, he studied psychology in the 1960s at Yale, where he did some early experiments, with exactly two rats, that explored an unusual dilemma in decision making: choosing between two unequal rewards separated by time, a smaller immediate reward pitted against a larger reward that took more time and effort to get. You don't need to be a rocket scientist to see the metaphoric import of such choices.

In order to test this, Ainslie first created a simple maze for rats to run through. Imagine you are a rat entering a doorway and facing two long corridors that run side by side. If you walked all the way down the corridor on the left (as Ainslie's rats were trained to do), curled around the corner at the far end, and walked all the way back, you would find your time and labors rewarded with a tasty little treat: three pellets of food. If, on the other hand, you walked to the right, you would see what appeared to be a similarly long corridor, but with a difference; there was a hole in the right-hand wall that let you get to the food right away, without having to walk all the way down the corridor and back. If you took this tempting shortcut, however, you got only one pellet. If the rats chose their corridor when they were near the hole in the wall, they picked the right side and got a small reward. This surprised no one: Researchers had earlier established that animals tend to undervalue (or "discount") larger but delayed rewards when offered smaller but immediate rewards, a process known as "temporal discounting."

But then Ainslie introduced a very clever wrinkle in the maze setup. He relocated the hole farther down the right-hand corridor, visible but not so close. This required the trained rats to make a choice before they reached the hole, and the animals showed a surprising tendency to prefer the left side (and its larger, delayed reward), avoiding the siren song of the hole in the wall that they would come to on the right side. This might sound like one of a million ho-hum rodent experiments, but it held the seed of destruction for a lot of models in classical economics, namely that preferences are constant and never change unless you get new information.

This avoidance of temptation was something that psychology—and economics—did not predict, but the whims of two rats were not enough to hang a publication, much less a theory, upon. In 1965, Ainslie had to drop his experiments temporarily to begin studies at Harvard Medical School. However, as a second-year med student, he dropped in on the laboratory of Richard J. Herrnstein, a behavioral psychologist most notorious to the nonscientific public for his coauthorship, with Charles Murray, of *The Bell Curve,* but much better known inside the scientific community (and, later, among behavioral economists) for his discovery of what is called "the matching law." In a famous series of experiments published shortly before Ainslie's arrival at Harvard, Herrnstein had demonstrated that animals that can get rewards from two side-by-side sources try them in proportion to how quickly, on average, they pay off. In other words, subjects' choices *match* the speed of the payoffs (as well as their frequencies and sizes).

These experiments did not seem to say much about the Ulysses problem at first, but Herrnstein set Ainslie up in the pigeon lab, and Ainslie tested the matching formula in birds, who faced a single choice between a larger, later reward and a smaller, sooner reward—the kind of choice he had given his rats. Herrnstein's formula predicted that an animal would prefer the smaller, sooner reward when close to it, but when farther away might be tempted to change its mind and choose the later, larger reward—the behavioral equivalent of tying oneself to the mast, as it were, in order to get the larger payoff. Some, though not all, of the pigeons learned to do this, and Ainslie plotted the preliminary results on a graph.

"I went to Dick," Ainslie recalled, "and said, 'Look, this is a hyperbola.'" What he meant was, the perceived value of the reward could be plotted as a curve on a graph—in fact, something called a hyperbolic discount curve, and you are invited to skip the following technical discussion of this graph, unless you've broken a diet, cheated on your spouse, awakened with a hangover, given in to a temptation, or otherwise done something on impulse that you've lived to regret. Yes, that means all of you.

Behavioral scientists had been aware of temporal discounting, but they'd assumed that both animals and people always discounted expected rewards in the same way that a banker discounts a future payment—by taking whatever its value is at a given time and subtracting a set percentage (the "discount rate") every time another unit of delay is added. (This seemed to make evolutionary sense, because it is the only way that relative preference between two rewards that you expected at different delays would stay the same as you got closer to them.) It would make the animal, or person, consistent over time, and therefore maybe more adaptive.

But hyperbolic discount curves are more simpleminded. They just take whatever value the reward has when it occurs and divide by the expected delay to get its current value. Hyperbolic discount curves predict weird distortions in the way we value things. An experience you can have a year from now—a day at the races, a night with your lover—will feel about as valuable as the same experience a year and a day from now. But an experience you can have tomorrow will feel twice as valuable as the same experience the day after tomorrow.

This is the same warp in perception that governs a lot of appearances, including how high a building looks as you walk toward it. Ainslie discovered that as you get closer to a small reward that is due before a larger alternative, at some point the earlier reward will feel bigger than the later one, just as a shorter building appears to loom higher than a taller one behind it as you walk closer to it. A hyperbolic curve describing the apparent size of the smaller reward, or building, spikes above the curve of the larger one as the smaller one gets close. You may still understand that the later reward, or building, is really bigger than the earlier one, but if the *feeling* has a direct influence on your behavior, like the song of a Siren, that understanding alone will not be enough to control your choice. Compared to the "exponential" curve that describes consistent preference over time, a hyperbolic curve is more bowed—like an archery bow pulled by its string. Because this shape tends to make your preferences inconsistent, it has the potential to do a great deal of mischief.

The reason this graph of our motivation—this chart, if you will, of what we *feel* to be most important—is so devastating to humans is because the trajectory of our desires and our decisions follows this same hyperbolic shape. In a large number of experiments that involve choosing between immediate and delayed rewards, human behavior consistently—and tragically—obeys this pattern. We cling to the scrawny bird near at hand and keep chasing it long after it should be clear to us that we should be focused on that much plumper pheasant in the bush. In our assessments of the value of rewards near and far, small and large, the two lines—which, again, reflect the way the brain attaches value to competing choices—cross because they are hyperbolic.

A less mathematical way of saying this is that we are neurologically doomed by this hyperbolic warp in our perception to see the smaller reward as bigger than it really is as we get closer to it, just as Odysseus would surely have been lured onto the rocks as he neared the Sirens. Even Socrates realized that this was a problem of a warp in perception: "Do not the same magnitudes appear larger to your sight when near, and smaller when at a distance?" Because he was talking about visual perception and scale, Socrates, without knowing it, was making a neurological point. In the idiom of modern architecture, Ainslie likens the paradox to "standing so close to a shorter building that you don't realize that the building behind it is much taller." So it is with the way our brains assign value to possible choices separated by time and size. If this isn't a dilemma for human wisdom, nothing is.

After Herrnstein set Ainslie up in the pigeon lab to repeat his discounting experiments, it took the better part of six years to gather all the data, but by the time Ainslie published the results—the first curves came out in 1974—the pigeon experiments improbably made clear why so many human choices are doomed to failure. Unless we somehow learn metaphorically to lash our impulses to the mast, short-term interests are destined to win out. These early experiments in the biology of patience and impulse caught the eye of economists, because the results suggested, inferentially, that consumers tend to impatience in the present and to patience when contemplating future decisions.

But these experiments suggest much more. As Ainslie wrote in *Breakdown of Will*, "The irony of smart people doing stupid things—or having to outsmart themselves in order not to—appears in literature again and again." And the key to overcoming the pressing urgency of immediate gratification requires the strength of which Saint Augustine often spoke: willpower. As Ainslie puts it, "It usually takes some kind of effort (willpower again) to evaluate a smaller present satisfaction as less desirable than a greater one in the future." When I spoke with Ainslie, he was not shy at all about linking the control of impulsivity to wisdom. "I would say the main purpose of wisdom," he said, "is to govern the will."

A clutch of familiar clichés flutters like moths around the flame of impatience: "the patient ant," "bird in hand," "haste makes waste," "hold your horses." All these expressions derive their moral juice from the tension inherent in decisions that force us to choose between a reward right in front of us (the bird in hand) and one that is larger, sometimes theoretical, and always in the future (two in the bush).

Patience is the battleground for now versus later, immediate reward versus delayed gratification, impulsivity versus prudence. These decisions are often life-defining, even though some are reached after much deliberation (whether to pursue an advanced degree, what kind of terms to accept on a mortgage, when to have children), while others are made on the spur of the moment (ordering another drink, breaking your diet, or choosing to have sex with someone, which can also, inadvertently, be about when you'll be having children). The element of time in human decision making, and whether to be patient or impatient, might be viewed as the fourth dimension of wisdom.

As off-putting as the jargon of hyperbolic and discount curves is, these studies on the neural correlates of patience focus on a constellation of human situations that separate wise behavior from foolish behavior: temptation, desire, lust, prudence, the pursuit and deferral of pleasure. Plato was among the first to cast this as a dustup between passion and reason, but of all the writers and philosophers who have waded into the psychology of this fray, perhaps none has metaphorically anticipated the underlying neuroscience better than Augustine of Hippo, the fifth-century saint, whose *Confessions* can be read as a young person's hell-raising catalog of temporal discounting, which Ainslie once described as a perfect formula for original sin and a form of "seduction by short-term rewards."

Augustine described his childhood as a succession of "mighty waves of temptation," and he gave in early and often. He told countless lies, stole from his parents, cheated at games, elevated the juvenile theft of a neighbor's pears to grand larceny, and, in young adulthood, applied the lessons of temporal discounting to lust and carnal excess. In a phrase that presciently links these serial transgressions to the brain's basic reward system, Augustine "became to myself a land of famine," which is a lovely way to frame impatience as the product of a relentless and insatiable hunger.

Augustine explicitly conceded to wisdom a leading role in his own version of this interior drama; he began to renounce "earthly things" at age eighteen after reading Cicero's *Hortensius*. "The love of wisdom has the Greek name 'philosophy,' " he wrote in Book III of the *Confessions,* "and it was with this name that that book set me ablaze." What he especially loved about Cicero, he explained, was that "his words aroused me and set me on fire not to be a lover of this or that sect, but of wisdom itself, whatever it may be; to love it and seek it and gain it and keep it, to embrace it with all my strength." (Note how the fervent syntax of this conquest, however, suggests that wisdom may merely have been the latest of Augustine's serial infatuations.) The most important word in that sentence is probably not *wisdom,* however, but *strength,* because it hints at a way around the problem of impulsivity.

We don't normally think of wisdom in terms of willpower, but Ainslie makes a provocative argument on this theme. He points out that classical Greek philosophers first characterized the conflict in self-control as a distinction between "wanting" and "judging." He wrote, "I *want* to have a love affair, but I *judge* it to be imprudent, or unethical, or sinful. How do I decide? No theory about how the wants and judgments compete for dominance has taken us beyond Plato's contest between passion and reason." But if we see this Platonic tension in neurological terms—with *wanting,* loosely speaking, associated with the immediate, possibly dopamine-driven emotional system of neural valuation, and *judging* associated with the cognitive, prefrontal, "future-seeing" part of the brain—Ainslie's notion of willpower becomes a potentially powerful player in this neural drama. In his view, individual willpower is a late addition to the conscious repertoire of human behavior, becoming prominent around the sixteenth century with the Protestant Reformation, but remaining mysterious as a process right down to the present. Within the framework of hyperbolic discounting, Ainslie sees willpower as a voice representing the future in the neural bargaining that goes on between the value of two choices, an interest in the debate between "the expected value of your future self-control against each of your successive temptations."

Almost every philosopher worth her or his contemplative salt has weighed in on this dilemma, from Socrates, who lamented on his deathbed the corrupting distractions of bodily gratifications, to Buddha, who laid out the five strategies for controlling the passions, to Saint Augustine, who conveniently converted, after a youth of impulsivity and miscreance, to the notion that "patience is companion of wisdom." But what does patience look like through the lens of neuroscience, and are there any cognitive muscles we can flex to make ourselves more patient?

We find clues not only in the work of epic poets like Homer, but also in that of enlightenment economists and philosophers. As was noted earlier in the chapter on moral judgment, Adam Smith recognized in the 1759 *Theory of Moral Sentiments* that the "passions" have special influence over decisions, including choices that pit immediate rewards against future ones. Two decades earlier, David Hume even offered a blurry sketch of the circuitry—neurophysiology on a philosophical napkin, you might say—in *A Treatise of Human Nature* when he asserted that reason, or rationality, "is and ought only to be the slave of the passions."

It's one thing to suggest a tug-of-war between passion and reason in the idiom of eighteenth-century philosophy; it's quite another to see impulsivity and delayed gratification duke it out in a brain scan. Not surprisingly, the neural aspect of this research arose in the field of behavioral economics, and has focused on economic decision making and choices in the marketplace. These are known as "secondary," or symbolic, rewards (such as money), as opposed to "primary," biological, rewards (like food and water), and so they may seem more narrow and constrained than the sorts of long-range decisions and choices that we customarily associate with wise behavior.

But scientists believe these economic choices are rooted in the same brain circuitry of reward that makes us feel so energized and motivated when we quench our thirst or sate our hunger (or, for that matter, learn something new). And these choices, therefore, may tell us a great deal about familiar human behaviors like procrastination and impulsivity; they clearly influence the way we plan for the future in terms of retirement savings, long-range investments, and credit-card borrowing. These are, in short, academic exercises that wallow in the messy reality of human imprudence.

Those experiments began to become possible one day in the mid-1980s, when Herrnstein invited Ainslie to give a lecture on impulsivity to his class at Harvard. Sitting in that class was an undergraduate economics student named David Laibson, a prodigy in math who had begun to wonder how you could mathematically adapt Ainslie's data about impatient pigeons to do experiments with people.

· · ·

It took nearly a decade, but by 1997, Laibson had worked out the math and published an influential paper in an economics journal on what he called the "quasi-hyperbolic time-discounting function." "It makes a lot of these concepts usable for economists," he explained, "and it's an easy way to translate these ideas in a way even a fifteen-year-old can understand." I could show you the actual math, but Laibson draws an even more vivid word picture of a formula in which "the present gets a special weight compared to all future periods." Thus, in a lot of situations that require wise judgment, the present appears to us like the Sirens on the rocks, while everything else appears smaller by comparison.

This isn't exactly a formula for wisdom, because it doesn't tell us how we *should* act. But it's done a pretty good job of explaining how we *do* act when faced with these kinds of decisions, in all its gory irrationality. In fact, it's been devastatingly accurate in predicting self-destructive behavior. With this formula, scientists could do experiments on human impulsivity, willpower, intertemporal bargaining, bad choices, falling prey to temptation—all the bad decisions that get us into all sorts of trouble. In short, it has allowed us to find our inner pigeon.

As an example of how it works, mathematically *and* psychologically, consider the decision whether or not to exercise. Let's say that all the benefits of exercise—a better-looking body, a thinner waistline, less risk of diabetes—could be assigned a value, such as eight units, while the *cost* of exercise is six units. "Most of the benefits of exercise are spread out over time and into the future," Laibson said, "but the costs of exercise are all borne at the moment of doing. Since the future gets weighted as one-half of the present, the future benefit is coded as four while the cost is six. Psychologically, this translates into a cost of six and a benefit of four. So am I going to exercise? No."

Laibson credits Ainslie with the critical insight. "George was the first psychologist to put all the pieces together," he told me. And he regarded it as an example of where "the social scientists were catching up to the lay public. Everybody understands that people have self-control problems. Today, I want to eat ice cream, and tomorrow I want to diet. Laypeople get that. They don't need to know that the hyperbolic discount curve produces that tendency." Neuroscientists needed to know that, however, in order to design experiments.

How does this feeble line on a graph translate into the workings of the brain and the myriad ways humans manage to be foolish? At a purely mechanistic level, that is what a team of researchers, including Laibson, set out to find. They couldn't put people through mazes when they were lying in MRI machines, but they could offer them choices that pitted present rewards against future ones while taking a peek at their brains. Their working hypothesis was that short-run impatience would be driven by the part of the brain rooted in emotion, which is more sensitive to immediate rewards and less sensitive to future rewards, while long-term patience would be controlled by the lateral prefrontal cortex and similar forebrain structures, which are better able to weigh the value of abstract rewards, including rewards in the future.

Imagine a situation where you are trying to discern the relative value of monetary payoffs instead of the relative height of buildings. You are offered a series of either/or choices about Amazon.com gift certificates. You are asked to choose between receiving a five-dollar gift certificate right away or, say, a thirty-dollar gift certificate in four weeks. Which represents the better value? Or twenty dollars now versus thirty dollars in two weeks? Or fifteen dollars now versus twenty dollars in four weeks? The calculus of evaluation that goes into each of these decisions is a little different. It changes with the relative value of the gift, the relative amount of time you might have to wait to collect your reward, and (though this isn't often emphasized by the scientists who conduct these studies) the personal tics we bring to the calculation of time and rewards, which vary, often significantly, from person to person.

Several years ago, Laibson and a group of colleagues asked Princeton University students to make precisely those choices about Amazon .com gift certificates while their brains were being scanned in an fMRI machine. The math underlying the experiment was Laibson's. The designated muse for this experiment was not Homer, however, but Aesop.

"In a field one summer's day," Aesop's famous fable begins, a grasshopper happily indulged in some invertebrate version of carpe diem, gamboling in the grass, living for the moment, longing to do nothing more ambitious than chat and sing. As he chirped "to his heart's content," one of those annoyingly selfless, industrious, eyes-on-the-prize ants labored by, arduously rolling an ear of corn back to its nest.

"Why not come and chat with me," the grasshopper said, "instead of toiling and moiling in that way?"

"I am helping to lay up food for the winter," replied the ant, "and recommend you do the same."

The grasshopper never bothered to lay up reserves for the winter. When winter inevitably came, he could only watch, perched on the edge of starvation, while the ants feasted on their banked stores of grain. And the moral (everybody, all together): "It is best to prepare for the days of necessity."

To a group of neuroscientists based at Princeton University, however, "The Ant and the Grasshopper" is the quintessential neuromorality tale about "temporal discounting." As they put it several years ago, "Human decision makers seem to be torn between an impulse to act like the indulgent grasshopper and an awareness that the patient ant often gets ahead in the long run."

The research team—led by Jonathan Cohen of Princeton and including Laibson, George Loewenstein of Carnegie Mellon University, and Samuel M. McClure, now at Stanford—originally intended to test their idea about temporal discounting by using fruit juice as a reward (thirst being a fundamental reward with deeper evolutionary roots than, say, a coupon for the latest Danielle Steel opus). While they were working out the logistical challenges of delivering a squirt of juice to someone lying prone in an MRI machine, however, it dawned on them that monetary rewards—in the form of gift certificates—could serve as a proxy (and the Princeton researchers, like many neuroeconomic empiricists, did not use Monopoly money in these experiments; the subjects received real gift certificates, so they had a real stake in their decisions).

Presented with a series of these either/or choices, subjects had to choose one or another Amazon certificate. Sometimes it was an immediate reward (today) versus a reward two weeks in the future; other choices involved taking either twenty dollars in two weeks, say, or forty dollars in six weeks. The results, according to lead author McClure and his colleagues, showed that the emotional, impassioned part of the brain—sometimes known as the "limbic system"—whipped into high gear when people made decisions about (or indeed were merely *offered*) immediately available rewards. It was, crudely put, the "instant gratification" part of the brain.

When the brain assessed all options, immediate and delayed, however, the researchers detected increased activity in what they considered a
more deliberative, cognitive part of the brain—specifically, regions of the
lateral prefrontal cortex and the posterior parietal cortex. This, put
crudely again, looked like the "delayed gratification" part of the brain.
The Princeton researchers also found that the subjects of the experiment
who really revved up the deliberative, cognitive machinery ended up
choosing the larger, more delayed options. When humans are forced to
choose between immediate and delayed rewards, they wrote in a widely
cited 2004 report in *Science,* "the idiosyncrasies of human preferences
seem to reflect a competition between the impetuous limbic grasshopper
and the provident prefrontal ant within each of us."

Whether that competition involves two separate decision-making
systems (as the Princeton-based group suggested) or simply one (as a
group from New York University, which published a strong counter-
argument in 2007, claimed) is still up in the air. Perhaps the more
important point is that the deliberative, rational, prefrontal part of the
brain plays a role—sooner or later, as consigliere, if not autocrat—in the
decision-making process when we try to sort out immediate versus
future rewards and exercise patience. Humans tend to be steep temporal
discounters (that is, when it comes to rewards, we much prefer sooner
rather than later), and yet these tendencies vary by age, personality, and
habit. Experiments have shown, in general, that younger people are
steeper discounters (that is, more impulsive) than older people, extro-
verts are steeper discounters than introverts, and drug addicts (and even
smokers) are steeper discounters than adults without such addictions.

Economists have long been fascinated by this behavior because of its
obvious implications for personal savings, public policy, and specific
future-oriented programs like Social Security. But just as clearly, this
kind of thought process has ramifications for life decisions that occur
outside the marketplace—choices that may involve educational plans,
careers, love affairs, job offers, or perhaps a response to a severe or even
terminal disease diagnosis. These are decisions that obviously have
rational economic components but also involve emotional impulses, so
that a "rational" (and wise) choice might be to select a job (or career) that
promises less income at first but will yield a larger income down the

road. These choices are further complicated by our struggle to attach abstract value ("priceless" considerations, as they say in the credit card ads) to a choice: considering, for example, jobs that promise a lower salary, but perhaps also less stress and more time spent with one's family. In the strictest sense of economic self-interest, choosing less income is always an irrational choice; if you expand the frame of evaluation to include emotional and familial "compensation" (something experimentalists normally don't do), some of these choices may seem more grounded and less irrational.

In 2007, the Princeton group published the results of a study testing the same idea with fruit juice instead of book certificates. Same result: Study subjects generally favored immediate smaller juice rewards, rather than a larger, though delayed, squirt of juice. And again, the emotional part of the brain responded most strongly to immediate rewards, while the cortical parts weighed in on the side of delayed gratification.

One of the criticisms of the 2004 *Science* paper is that its rendering of a Hume-like cerebral brawl between emotion and reason is "way too simplistic," in the words of Stephen Kosslyn of Harvard University, who specializes in the brain's visual system but has been an avid sideline observer of social neuroscience. Researchers at NYU have been more explicitly critical; they say their results "falsify" the Princeton hypothesis, claiming the evidence is unambiguous that human decision making of this sort represents a single continuum of evaluation and choice. But there will probably be a few more twists in this story, and as Kosslyn pointed out, "There's no complex behavior that's under the purview of one system [in the brain], so I don't know why one as complicated as this one wouldn't have more than one system." Ainslie, writing about the Princeton experiment in *Science* with his colleague John Monterosso, noted, "This work is an important step toward direct observation of the decision-making process, although its findings are open to different interpretations."

The biology of patience, and impatience, is far from settled. Another avenue to a biological understanding of patience comes to us, surprisingly, from the medical treatment of people with Parkinson's disease. These patients suffer from a diminished production of dopamine in a part of the brain known as the subthalamic nucleus (STN), a structure deep in the center of the limbic system. As a result, Parkinson's patients suffer motor dysfunctions, including muscular rigidity and uncontrollable trembling. A recent new treatment is a technique called "deep brain stimulation," in which surgeons delicately implant electrodes near the subthalamic nucleus; later, a pacemaker-like battery sends to that spot pulses of electric current, which seem to alleviate some symptoms.

But doctors have recently observed a bizarre cognitive side effect to deep brain stimulation: It appears to make some Parkinson's patients behave impulsively. They will rashly decide to do something—walk across a room to a more comfortable chair, for example—and promptly fall down. Other Parkinson's patients, treated with so-called dopamine agonists (drugs that promote the release of dopamine), have similarly reported a loss of impulse control, including episodes of pathological gambling and hypersexuality.

Intrigued by these odd side effects, neuroscientist Michael J. Frank and his colleagues at the University of Arizona tested the decision-making behavior of Parkinson's patients against a similarly aged control group, and they recently reported that deep brain stimulation interferes with a natural tendency to slow down when a person is confronted by decisional conflict. Instead, these patients jump to conclusions.

Frank's theory, which is still being tested, "predicts that when faced with multiple seemingly good options, the subthalamic nucleus enables you to adaptively 'hold your horses,' buying more time to settle on the best one." In other words, he believes there's something in the STN, deep in the mid-brain, that imposes patience on a decision when the brain detects conflict. "I think of it as a basic mechanism that allows a little more time to make a decision," Frank told me. "It's possible that the same mechanism might prevent you from saying rash things in a family environment. But once you get into that kind of situation, there are so many other things in play that it would be hard to model a mechanism that accounts for that much complexity. But that kind of domestic situation may be tipping into these basic mechanisms."

Like moral judgment and compassion, patience may ultimately have a long evolutionary history, so researchers have even begun looking for answers in a population not usually known for impulse control: animals. Over the past five years, scientists at the Max Planck Institute for Evolutionary Anthropology, in Leipzig, Germany, and at Harvard University have published several articles on the evolution of patience in monkeys, chimpanzees, bonobos, and humans. Recent research, led by Alexandra G. Rosati and published in 2007 in *Current Biology*, reports that both bonobos and chimps demonstrate a degree of patience not seen in other animals. Indeed, some primates exhibit degrees of patience not found even in humans! In comparative experiments between primates and humans, we proved to be the impulsive ones, "less willing to wait for food rewards"—that is, more impatient when it comes to grub—"than are chimpanzees." The researchers also reported that "humans are more willing to wait for monetary rewards than for food, and show the highest degree of patience only in response to decisions about money involving low opportunity costs."

It is hard to exaggerate the implications of something as obscure as the hyperbolic discount curve for human behavior and indeed human destiny. As economists have pointed out, some of the best-known examples of long-range planning by humans—making provisions to leave money to grandchildren, for example, or acting to preserve the environment, or socking away cash in a savings account—obey the hyperbolic curve, which doesn't bode well for grandchildren, the environment, and the retirement plans of baby boomers.

The curve captures the way we give in to temptation when we abandon a diet and impulsively pig out; it captures the way we procrastinate and put off doing what needs to be done. It even captures, by extension, the cruel cycle of behavior that Augustine elevated to the timeless literature of self-loathing fifteen hundred years ago and that has been keeping confessors, clergy, and psychotherapists in business ever since: temptation, followed by impulsive giving in, followed by enthusiastic surrender and pleasure (all registered by the brain's dopamine system, its neurochemical embers glowing in the hearth of gratification), followed inevitably by regret and self-contempt. The future disappears in a fog of irresistible desire and inevitable surrender.

But as both Odysseus and the ant make clear, the future is where the action is when it comes to wisdom, and no one has conceptualized the psychological and cognitive importance of this better than Ainslie. The only way to offset the allure of the Sirens is to nurture some idea of the future—a cognitive act of imagination sometimes verging, he admits, on fantasy—that seems more rewarding than the present. This places an inordinate amount of pressure on the human imagination, but that's one of the most appealing aspects of Ainslie's ideas about willpower.

In the marketplace of the brain, he believes, this boils down to understanding that what you do in the present when confronted with temptation predicts all your future behavior. If you understand that, if you "bundle" the next cigarette or the next drink with the desire for who you want to be in the more distant future, you have a chance of overriding the immediate desire. Every time you forgive yourself by saying, Just this one time, you are sabotaging your future self. Every time you give in to impulse, you subtly erode your chances of becoming the person you hoped to be in the future. Every time you bet against your future self, you increase the odds of repeating the regrettable behavior again and again and again. No wonder, then, that Ainslie thinks of wisdom as a way of strategizing about the future.

"The only reason we're not pure calculating machines," he says, "is the hyperbolic shape of the discount curve, which means you have to have a strategy for long-term rewards." To those who face temptation of any sort—sex, booze, a slice of pie, an excuse not to do homework—Ainslie believes, wise behavior boils down to one thing: "Wisdom as it applies to self-control is really the awareness that what you do now predicts your future." But, in keeping with the contextual nature of wisdom, Ainslie's version of self-control is not an either/or, on/off switch. "You can have too much self-control," he said. "A ham-handed concern with how every choice predicts your future can make you compulsive. So the highest wisdom is the art of balance."

At scientific meetings these days, George Ainslie likes to show several facetious slides that attempt to recast this age-old Platonic dilemma in the idiom of Aesop. One plays off the McClure theory: It shows an ant and a grasshopper tussling over the steering wheel of a moving car, fighting to assert control. The other shows an ant riding on the back of the grasshopper, dangling food on a pole in front of the grasshopper in order to influence its desires and motivations. He believes that impulsivity, in the form of the grasshopper, drives the brain, and that prudence about the future, in the form of the ant, has to ride this impulsivity like a horse. The farsighted ant, in Ainslie's view, can muster influence only by what it shows the grasshopper; by being strategic and imaginative in its trickery, such as bundling together a believable path of rewards into the future, it asserts and steers behavior. In Ainslie's view, the ant (the cognitive part of the brain) cannot control the emotional core of impulsivity and impatience that drives so much of our behavior; but, with a little creativity and understanding of the way the brain responds to reward, the ant might be able to steer the more powerful beast upon which it rides.

CHAPTER ELEVEN

DEALING WITH UNCERTAINTY

*Change, "Meta-Wisdom," and the Vulcanization
of the Human Brain*

> *Whoever cannot seek the unforeseen sees nothing,
> for the known way is an impasse.*
>
> —Heraclitus

> *The world we have made, as a result of the think-
> ing we have done thus far, creates problems we
> cannot solve at the same level of thinking at which
> we created them.*
>
> —Albert Einstein

JONATHAN COHEN, who directs the Center for Brain, Mind, and Behavior at Princeton University, knew he was making an outrageous
rhetorical point over lunch one day, but he clearly liked the sound of
it. Hunched over a bacon, lettuce, and tomato sandwich in a Princeton
luncheonette, Cohen wore a gray T-shirt, blue jeans with rips at the
knees, dark eyebrows behind Harry Potter glasses, and a shock of star-
tlingly white hair that fairly waved in the conversational breeze as his
thoughts careered down alleyways at high speed, with lots of gear shifts,
lane changes, and precious little braking for turns. "I bet I could beat
him," he said.

Cohen was making a point, without explicitly saying so, about the
difference between intelligence and wisdom. He was explaining—well,
speculating, really—about how a well-trained and expert, but inflexible,

mind can get tripped up, behaviorally and *neurologically,* by a small change in the rules by which it typically makes decisions and solves problems. His little lunchtime fantasy arose from this bold prediction: If you changed a single simple rule in chess—making the knight move up *two* squares instead of one, for example, before making its diagonal jump—it would so flummox a grand master like Garry Kasparov that Cohen (a self-described duffer in chess) could adapt more quickly to the change and defeat the former world champion.

"I bet I could beat him," Cohen said as he took a bite of his sandwich. As he warmed to the outrageousness of the thought, he kept repeating it. "I bet I could beat him. You know, I bet I really *could* beat him."

There were two interesting things going on during that lunch, both of which may bear not only on the way wisdom percolates through the brain, but also on how we think about the way that the mind works as it works its way through an unfamiliar problem. The first had to do with eating that BLT while following the thread of our conversation. In the annals of human cognitive achievement, keeping a firm grip on a deli sandwich and eventually making it disappear does not exactly rank as a breathtaking accomplishment. Yet, having eaten his share of sandwiches over the years, Cohen did not need to spend a lot of time thinking about each separate and significant decision that goes into such a task: maneuvering the sandwich smoothly to his open mouth, taking a bite of more or less the right size, chewing the food down to a malleable texture so that it would squeeze through the esophagus, swallowing, and making sure not to breathe (or laugh) as he swallowed—all while conducting a conversation about wisdom, chess, and the brain. The automaticity of all those actions, the utter *thoughtlessness* that went into each sequential decision, allowed Cohen's mind to do other things at the same time: listen to a writer blurt out a barely coherent question, try to make sense of what he was trying to ask, discern the right level of jargon with which to respond, think through an answer, and begin to deliver it, being careful not to speak with his mouth full.

That may not sound very cognitively challenging, either, but Cohen was making a point that is very relevant to how we deal with uncertainty, which is a cornerstone to all modern psychological definitions of wisdom. He believes the human mind deals with multiple tasks and decisions by mobilizing multiple systems of thought. One is essentially automatic and conforms to the world—of sandwiches and virtually everything else—as we've always known it. This is widely known as "model-based learning," according to Cohen, although we might simply think of it as experience: We compare a possible future action or choice with similar situations we've confronted in the past.

The other way (or at least *another* way) of negotiating the world is what neuroscientists call "model-free" learning. This occurs when we are confronted with something novel or unexpected, something that doesn't align with our prior experience. We need to pause to gather more information and evaluate different options prior to making our decision. Model-free learning is obviously much less automatic, and consequently takes more time; it probably represents the kind of situation that Euripides had in mind when he said, "In this world second thoughts, it seems, are best." Uncertainty, change, the unexpected—these all demand a different kind of appraisal, a different process of evaluation. Which brings us back to Kasparov and the world of the chessboard.

Grand master chess players, Cohen said, think ahead from situations to outcomes in an automatic fashion. We don't. A model-based approach to a chess move, based on a boatload of previous experience, allows an expert to make decisions quickly and efficiently, but without much flexibility. Change the rules, however, and suddenly the entire data set of "knowledge" is corrupted by the worm of novelty. When history repeats itself, in other words, the model-based neurological approach to processing information and dealing with a problem is superior—indeed, not only contributes to efficient and correct decision making that might be considered wise but, in certain situations of threat, is probably lifesaving. When history *doesn't* repeat itself, and you need to adapt to a different set of circumstances, the wise course of neural action is to be flexible, drop the embrace of old rules (and old habits), and reevaluate all the information.

To a certain extent, Cohen probably exaggerated the automatic nature of chess decisions to make his point. In his recent book *How Life Imitates Chess,* Kasparov, in fact, describes a process in which the ability to adapt is crucial to success. "A Grandmaster," he writes, "makes the best moves because they are based on what he wants the board to look like ten or twenty moves in the future. This doesn't require the calculation of countless twenty-move variations. He evaluates where his fortunes lie in the position and establishes objectives. Then he works out the step-by-step moves to accomplish those aims." Kasparov makes it all sound logical, flexible, and shrewd. But later he admits that his true gifts in chess stem from memorization as well as "fast and deep calculation"— both qualities associated with a kind of automatic mental processing. We might even think of it (because brain scientists already do) as a process that triggers algorithms embedded in the neuronal structure of our gray matter.

The problem, of course, is that life (and wisdom) isn't like chess. John von Neumann, one of the grandfathers of game theory, made precisely this point when he called chess "a well-defined form of computation. . . . Now real games are not like that at all. Real life is not like that."

Real life is not confined to a board; it is not symmetrical and neat; it does not always reward boldness and aggression; it does not forbid a new piece from flying in out of left field. Sometimes the rules change abruptly and without warning. Sometimes the situation looks deceptively similar to a previous scenario but isn't. Sometimes you are absolutely certain about something, but you turn out to be absolutely wrong. One of life's most daunting challenges lies in the fact that if you try to plan twenty moves ahead, you may find that the situation, the rules, indeed all the things you assumed to be stable and predictable about your world, have changed dramatically halfway to your goal—an illness in the family, the loss of a job, an unexpected failure, not to mention age-old calamities of biblical proportion, like floods, hurricanes, fires, and plagues. We don't know if Heraclitus played chess, but we do know that he understood the game of life. "All is flux," he said. "Nothing is stationary."

The point Cohen was making, in this lunchtime fantasy about Kasparov, was ultimately one about what he called, very much off the cuff, "meta-wisdom." It amounted to a higher level of understanding, almost an intuition, of knowing when experience should guide decisions and when you have to throw out the experiential playbook and literally do a rethink. In this highly specialized form of understanding, you would have an intuitive feel for knowing which decision-making system served you best. In other words, wisdom is not simply a matter of knowing the best answer to a problem or dilemma; it is a matter of knowing the best *approach* for finding that best answer.

· · ·

On the brambled path to worldly wisdom, declarations of certainty often meet hidden patches of quicksand along the way, and even the most imposing intellects stumble into these traps. G. E. Moore, the influential twentieth-century English philosopher, made precisely such a misstep in a famous (among philosophers) talk he gave in 1941 at the University of California at Berkeley about certitude. Moore pointed to the ceiling of Wheeler Auditorium and declared, as an example of the kind of thing that humans could know without a doubt, that the light streaming through the recessed windows came from the sun. "Most in the audience were aware, however, that the glass panels were diffusers for electrical illumination; the roof of the building was solid and opaque," recalls philosopher Wallace Matson in *Uncorrected Papers*. When someone in the audience pointed this out, Moore, one of the founders of analytical philosophy, blurted, "Oh dear me!" and moved on to the next question.

As Matson notes, in an essay whimsically entitled "Certainty Made Simple," certainty has more to do with people and their belief systems than with any given set of facts. We need the comfort (and, often, illusion) of certainty to move on to the next question (or dilemma, or crisis) in our lives, even though we are buffeted by uncertainty and flux. Knowledge offers comfort; wisdom feels comfortable with uncertainty. Making a distinction that echoes all the way back to ancient Greece, Rabelais famously reminded us, "The greatest scholars are not the wisest men."

Psychological scholars of wisdom—everyone from Paul Baltes to James Birren to Monika Ardelt—have repeatedly emphasized the point that the ability to deal with uncertainty, both emotionally and intellectually, is a crucial aspect of wise behavior. But how do our brains actually react when we are confronted with novelty and uncertainty? According to Cohen, we deal with them by struggling to reconcile the implorations of the emotional brain, which sets up swift, automatic responses (and, in turn, allowed humans to survive in the evolutionary past), with the cognitive brain (which will have a lot to say about how long we will survive in the future). A blink of wisdom is not likely to help us sort out this problem. More like a pause, but not too long a one. When Confucius learned that one of his disciples weighed each decision three times before acting, the Master disapproved: "Twice is quite enough." Confucius believed, according to later interpretation (and in direct contradiction of modern economic theory), that "if one thinks more than twice, self-interest begins to come into play."

Adjusting to the unbearable persistence of uncertainty may be one of the loftiest accomplishments of human wisdom. In a beautiful passage from Book V of his *Meditations,* the Roman emperor Marcus Aurelius—known in the Roman era as "the wise one"—captured something of the contingent nature of life and how we might respond to it. "Repeatedly, dwell on the swiftness of the passage and departure of things that are and of things that come to be. For substance is like a river in perpetual flux, its activities are in continuous changes, and its causes in myriad varieties, and there is scarce anything which stands still, even what is near at hand; dwell, too, on the infinite gulf of the past and the future, in which all things vanish away. Then how is he not a fool who in all this is puffed up or distressed or takes it hardly, as if he were in some lasting scene, which has troubled him for long." Only a fool, in short, would be upset by change.

How do we make decisions (wise or otherwise) in the face of this perpetual flux? We often have a conversation in our heads between intuition and thought, emotion and reason, a hypothesis (or idea) about what is real and then a burst of analytical thinking to see if what we suspect is true actually is. At its most rigorous and efficient, this internal dialogue under the hood resembles the version of scientific reasoning articulated by the essayist Peter Medawar.

In discussing the role of intuition in scientific reasoning in *Pluto's Republic,* Medawar focuses on the back-and-forth between hypothesis and testing, intuition and critical thinking, which, in a way, captures the difficulty and also the power of "meta-wisdom" that Cohen talks about. Medawar describes the scientific method as "a potentiation of common sense," and adds, "Like other exploratory processes, it can be resolved into a dialogue between fact and fancy, the actual and the possible; between what could be true and what is in fact the case." It is the process of going back and forth between hypothesis and reality, finding the program that "matches" the circumstances. Neuroscientists sometimes refer to this process as "framing," or perspective taking. Scientists are fond of saying, "If you don't formulate the question correctly, you will never get the right answer." They are typically referring to an experimental question, but the point is just as salient—indeed, more so—in everyday problem solving. The way we frame a problem predicts the success with which we work our way toward a solution.

Another way of thinking about framing is to understand that wisdom is contextual. There is a simple but spectacularly revealing example of this in Book XI of the *Analects,* where Confucius gives two different answers to the exact same question. Tzu-lu approaches and asks the Master, "When one hears a maxim, should one at once seek occasion to put it into practice?" Confucius replies, "Your father and elder brother are alive. How can you whenever you hear a maxim at once put it into practice?" Jan Ch'iu then asks the very same question, and Confucius replies, "When one hears it, one should at once put it into practice." The Master is immediately asked why he gave such starkly different answers to the same question, and he replies, "Ch'iu is backward; so I urged him on. Yu is fanatical about Goodness; so I held him back." It's not just how the question is framed, according to Confucian wisdom, but who is asking it, and how prepared they are to receive possible answers.

Meta-wisdom is an idea that lies at the heart of complex judgment and decision making. How do we assess the information we've learned (from book learning, from work experience, from episodes of heartbreak and savvy), and how do we distinguish a novel, uncertain set of circumstances from a standard situation where resolution lies in past experience? Put simply, how do we deal with uncertainty and still make the right decision? This problem is internalized as a conflict in the brain, Cohen believes, but the conflicted brain also holds the answer to the dilemma.

Meta-wisdom is neither a scientific nor a philosophic term; it was the small change in a conversation over lunch. But it struck me as an immensely useful metaphor for some of the more emergent qualities of wisdom, and it came from one of the more global and adventurous thinkers in the field of social neuroscience. Jonathan Cohen was the senior author on several key experiments we've discussed—Josh Greene's work on moral judgment, Sam McClure's work on delayed gratification, Alan Sanfey's view of the brain as it plays the Ultimatium Game. He likes to speculate about the implications of current brain research for dealing with climate change or future policy decisions. He likes to push ideas to their deepest, and sometimes direst, destinations—like the sobering political implications of our automatic, emotional decision-making apparatus in a future nuclear confrontation, for example. And he likes to have a little fun doing it, making frequent allusions to popular culture in his papers, borrowing from *Star Trek* here, quoting lyrics from Bob Dylan there.

Cohen's notions about meta-wisdom grow out of a largely unfunny argument he and others have been having, in the narrow field of neuro-economics and the larger field of cognitive processing, about whether decision making is a universal process, with a single universal neural mechanism for action, or whether the mind works with a multitude of processes, often in relative harmony, sometimes in conflict, almost always involving a competition between the older, innermost, emotional part of the brain and the newer, outermost layers of neural tissue known as the neocortex. We have known for a number of years, thanks in part to the pioneering research of Antonio Damasio and Joseph LeDoux, that a surprising amount of decision making is driven by the unconscious, emotional mind, and we also know that the cross-traffic between the cortex and the older, emotional brain architecture is so extensive, so interweaved and connected, that it's almost misleading to speak of two separate entities. We spend a lot less time thinking about when these emotional prompts are good and when they are not, when they're superable and when they're inflexible, when they inform wise judgment and when they ordain bad behavior. How could this scientific debate *not* offer profound insights about modern notions of wisdom?

Cohen has had a hand in two recent scientific papers that, to my mind, have important implications for our nascent neural notions of wisdom, although the word never crops up in either paper. One is a 2005 essay entitled "The Vulcanization of the Human Brain," in which Cohen describes the conflicted decision making that occurs when the emotional and cognitive parts of the brain urge divergent courses of action—an updated retelling of the old passion-versus-reason tale, and one he thinks could have ominous implications for public policy. The other is a 2007 paper by Cohen and two colleagues, entitled "Should I Stay or Should I Go?" Inspired by the Clash song of the same name, it specifically discusses how the brain deals with the Heraclitean dilemma of uncertainty and flux.

.   .   .

In *One Minute to Midnight,* a harrowing historical retelling of the 1962 Cuban missile crisis, Michael Dobbs documents a series of peripheral decision points that brought the world to the brink of nuclear war. A general theme of recent popular accounts of neuroscience has been to extol the "wisdom" of the emotional brain, but my initial reaction after reading Dobbs's account was, Thank God for the prefrontal cortex.

On several occasions as the crisis deepened, cooler heads succeeded in tamping down emotional impulses that, if acted upon, could easily have plunged the world into nuclear holocaust. On October 27, 1962, when an agitated and paranoid Fidel Castro repeatedly urged his Soviet allies to initiate acts of aggression against the United States (including a pre-emptive nuclear strike), the Soviet ambassador to Cuba, understanding the risks of those provocations, defused Castro's anger. Dobbs reports that on that same "Black Saturday," after American naval forces had flushed a Soviet submarine to the surface of the Atlantic Ocean south of Bermuda with "practice" depth charges, the outraged Russian commander prepared to retaliate with a ten-kiloton nuclear torpedo against U.S. ships in the vicinity; his junior officers "persuaded him to calm down."

The U.S. president John F. Kennedy emerges from these pages not only as preternaturally even-tempered and "prefrontal"; Dobbs describes him as having some of the qualities that psychologists have recently associated with wisdom, including an "introspective, skeptical nature" shaped in part by ill health during childhood and adversity during young adulthood (notably his close call with death during World War II). He was "forever questioning conventional wisdom" and "had a knack for looking at problems through the eyes of his adversaries." Geopolitical conflicts are not extraneous to neuroscience; as Cohen suggests in "The Vulcanization of the Human Brain," "The evolution of our emotional apparatus did not anticipate a world in which aggression can be expressed impersonally over large distances."

The basic argument of Cohen's "Vulcanization" theory departs from an evolutionary understanding of the brain that is hardly new and indeed has become clear through decades of painstaking animal research. The "old brain," in this view, is a tight, fist-size core of neural structures centered in the subcortical forebrain, with some extensions into the midbrain (Cohen refers to it as the "limbic system," although some neuroscientists find this term archaic and misleading). This is the classic "emotional brain," driven by basic appetites and physiological needs, orchestrated by a neurotransmitter (dopamine) that registers rewards and reinforces optimal behavioral choices. The "new" part of the brain—the neocortex—is literally an evolutionary afterthought in humans, a blanket of neuron-rich tissue, thin as a loincloth, draped over the old brain. Yet this frail three-millimeter-thick, six-ply tissue of neurons is responsible for all our higher cognitive functions: planning, abstract thinking, decision making, and considering the future consequences of actions. When predicting the geography of brain regions, we often think of front and back, top and bottom, left and right, but I've come around to the view that an equally useful distinction is to think of inner and outer, or core and periphery. The old, limbic, emotional brain is the lumpy, amalgamated ball of neural tissue bunched around and above the brain stem; the neocortex is spread thinly over the margins, nestling over the old brain almost like a shower cap.

The kernel of Jon Cohen's argument is that while these two systems often work in concert, there are times when they are in conflict. Indeed, he cites examples from several recent avenues of research we've already discussed—moral reasoning, with its conflict between emotional and utilitarian decisions, and altruism, with its conflict between individual and communal interests—as evidence of these neurological tensions. They arise, he believes, because the world has changed dramatically, socially and environmentally, over recent evolutionary time, while the emotional brain has not. This disparity accounts for all the so-called irrational behaviors that behavioral economists have giddily documented in recent years—bad choices about delayed rewards (driven by the emotional part of the brain), addictive decisions that reflect impulsivity, and so on. As Cohen puts it, "a broad range of decisions engage evolutionarily old brain mechanisms that have consistently been implicated in emotional processing."

The emotional brain is rapid, stereotypical, and inflexible. The newer part of the brain, especially the prefrontal cortex, is usually slower and more limited, but, paradoxically, more nimble when faced with the unexpected. Cohen believes it has the ability to align our thoughts and actions with abstract goals, and is particularly important when we have to overcome "countervailing habits or reflexes." In other words, Cohen, too, sees the cognitive part of the brain as a rider; he just equips it with better stirrups and a sharper whip in dealing with the emotional horse than does someone like Haidt. The question is, Is that enough to keep us out of trouble?

"Vulcanization," Cohen goes on to write, "is the process of treating a substance (such as rubber) to improve its strength, resiliency and usefulness. Similarly, evolution seems to have vulcanized the human brain through the development of the prefrontal cortex." This cortical shrink wrap, hugging the curves and crevices of the brain, has a number of functions: It offers general reasoning ability, it creates the kind of cognitive "tricks" George Ainslie has so deftly cataloged to overcome the impulses of the emotional brain, and it might even protect us from our ancient evolutionary impulses. As Cohen points out, neuroeconomic studies have shown that activity in the prefrontal cortex is associated with decisions that override pride or impatience. Moreover, this new part of the brain seems to be involved in the kind of *systematic* social activity and training designed to overcome the short-term urges of the emotional brain. The kind of thing we might think of as civilization— or institutional wisdom.

Consider the training of doctors and soldiers, for example. Their education is intended in part to overcome the strong (and quite reasonable) emotional aversion most of us instinctively feel when we witness suffering, severe injury, or death in others; no experiment to date has shown that overcoming those aversions requires the prefrontal cortex, but as Cohen shrewdly points out, the conception and design of training programs to overcome those aversions surely involved the prefrontal cortex. Similarly, we have devised financial policies and technological innovations to counteract the innate human vulnerability to instant gratification and addictive behaviors, including everything from Social Security and 401(k) plans to nicotine gum and Antabuse for alcoholics. "Those measures are clearly designed to "protect us against ourselves," Cohen writes, echoing Ainslie's riff on Ulysses and the Sirens, and, like the training of doctors and soldiers, required use of the prefrontal cortex to design and execute the plans. Institutional and social wisdom, in this view, creates systems and even vast governmental bureaucracies to outwit the quicker, self-defeating impulses of the emotional brain.

These prefrontal abilities have implications for everything from warfare to the way we respond to advertising, and echo many of the points George Ainslie makes about the way most people continue to discount the future when they make rash decisions. "Steep discounting may have been highly adaptive when most (if not all) valuable resources were perishable or were difficult to defend given the lack of well-defined and well-enforced property rights," Cohen says. "However, with the evolution of the prefrontal cortex, and the concomitant development of technologies from refrigerators to bank accounts, steep discounting and impulsive behaviors are substantially less adaptive, yet they persist." Moreover, he argues that prefrontal thinking has also been deployed to more nefarious ends, such as producing a bewilderingly large array of desirable material goods as well as concocting the clever advertising and marketing schemes that specifically target the impulsive part of the consumer brain. As a result, people are constantly induced to spend more than they have and to save less than they need. (In a similar vein, Ainslie attributes a good part of the recent financial fiasco to mortgage inducements that appealed precisely to this ancient and impulsive part of the brain.)

Not that we need anything else to worry about these days, but Cohen sees potential peril in the disconnect between the speed of modern innovation and the traditional pace of human evolution. The recent emergence of the prefrontal cortex, along with all the technological and social innovations it has spawned, has wrought environmental change at a rate that has outstripped evolution, creating situations and dilemmas for which the emotional brain can provide very bad counsel. Coupled with the fact that human genetic diversity suggests a vast, and probably disquieting, range of prefrontal development in different individuals, Cohen fears "a potentially fundamental instability: A world in which the potency of technology introduced by the prefrontal cortex, and perhaps manageable by it, is equally accessible to mechanisms that were not adapted to the use of that technology." (As an aside on the role of wisdom in exercising military potency: Has there been a major leader in modern times more openly, indeed boastfully, proud of "gut" decision making, which relies on the emotional part of the brain, than George W. Bush?)

Cohen's ideas about the "prefrontal brain" sound provocative and interesting, but in Peter Medawar's important litmus test, do they add up to a true story about neural life? Many neuroscientists remain skeptical, some vigorously so.

But there is some evidence for multiple systems for decision making, and it has implications—albeit still premature ones—for wisdom. Peter Dayan of University College London and his colleagues have identified in, if you'll forgive the phrase, "wise mice" at least four distinct neurological approaches to problem solving. The problem these rodents solve, of course, is quite modest: navigating a maze, and then making a choice between food or liquid. If the situation is "stationary" (that is, if nothing in the environment changes from one trial to the next, and the reward can always be found in the same predictable place), the animals rely on a pure algorithm of "reward-seeking," a fast and automatic series of decisions to navigate the maze. The advantage of this approach is its speed and efficiency; the downside, research is beginning to show, is that once a behavior becomes ingrained as neurological habit, it is extremely resistant neurologically to change, even when circumstances shift. By contrast, when environmental circumstances change (the food is hidden in a different place, or the liquid is relocated), the mice have to "reason" through a stepwise set of multiple decisions, each one leading to another decision point, to relocate the reward. This process is obviously slower, although more flexible. This trade-off between automaticity and flexibility is one that probably applies to humans, too.

We can leave it to the neuroscientists to decide whether two separate neural systems are fighting over the control switch of these decisions, or whether two different subsystems funnel competing information into a central evaluator. The more important point for our understanding of wisdom, especially as it guides us to lead a productive everyday existence, is that both systems are neurologically plastic (that is, permissive to change), both seem to change with age, and, to the extent that we are willing to be more effortful than mere unconscious intuition would imply, both seem to suggest that repetition or training, if you will, can make cognitive performance swifter, stronger, and maybe, just maybe, a little wiser.

Not that this settles anything. Confucius insisted, "He that is really wise is never perplexed." Socrates, Karl Jaspers reminds us, made self-inflicted perplexity the highest form of wisdom.

· · ·

Balance is a persistent theme in the literature of wisdom. The Buddha counsels it, and Confucius made the point that too much is just as bad as not enough. One of the qualities Marcus Aurelius most admired about his father was his (appropriately enough) horsemanship—specifically "the experience that knew where to tighten the rein, where to relax." Is there a neurological version of balance, of an Aurelian feel for what to do?

Something like neurological balance is beginning to attract scientific attention. What happens in the brain when higher organisms, including humans, must make decisions in the face of change? In strictly scientific terms, that is the question that Jon Cohen and two colleagues, Samuel McClure and Angela Yu, address in "Should I Stay or Should I Go?" In a broader metaphoric sense, they are talking about nothing less than the everyday dilemma of relinquishing comfortable habits and searching for better alternative behaviors when we suddenly find the world changing around us.

Cohen and his colleagues couch this as a choice between "exploitation" (staying with a winning behavioral strategy that satisfies one's needs) and "exploration" (testing other alternatives when our habitual strategy stops delivering the goods). Put another way, exploitation reflects automatic habit, while exploration reflects tentative adjustment to change. "The need to balance exploitation with exploration," they write, "is confronted at all levels of behavior and time-scales of decision making from deciding what to do next in the day to planning a career path." In terms of human decision making, it doesn't get more global than that.

It's a dense paper with elaborate formulas for behavior and a lot of technical neurophysiology. I'll skip the details, in part because they don't yet add up to a single coherent message; one of the conclusions of the paper, in fact, is that this is such a dauntingly complex subject to tackle, and includes so many real-life variables, that it may never be possible to model neurologically with anything other than cartoonish simplification. In a perverse way, however, that is why this little branch of research closely approximates experimental approaches to the subject of wisdom, and probably why these questions have attracted some of the best and brightest researchers in the field of social neuroscience—not just Cohen but also Peter Dayan in London, Yael Niv at Princeton, Liz Phelps and Nathanial Daw at NYU, Read Montague at Baylor, and others. You don't have to be a neuroscientist to be excited by the enormous philosophical question embedded in the nitty-gritty of this relatively brief paper. And for people who want wisdom to retain a bit of its mystery, we may even be glimpsing a wall beyond which science will never be able to peer.

First of all, like moral judgment and compassion and emotional regulation, the story is becoming increasingly vexed by complexity. How exactly, for example, do we convince ourselves to abandon a bad habit? We have heard a good deal about the dopamine reward system, but Cohen and his colleagues review some recent evidence that proposes an important role in habit breaking for two other well-known neurotransmitters, acetylcholine and norepinephrine. These molecules appear to act as sentinels, signaling uncertainty to the brain. They may even be implicated in the neural origins of boredom.

But at root, this is a paper about balance. It describes how people neurologically weigh the relative merit of sticking with a behavioral strategy or changing, of understanding when an immediate reward is right and when a long-term payoff might be greater, all as a product of adapting to a changing, unpredictable, or, as the authors put it, "nonstationary" environment. At a party, in a marriage, at a job, in a stock fund, the question is always the same: Should I stay or should I go?

Does all this dense, constrained, hyperqualified, and speculative science-speak ultimately tell us anything useful about wisdom?

Well, yes. First, it tells us that the brain probably acts differently when it is confronted with uncertainty or change. As a result, we need to be open-minded, in the metaphoric but also neural sense. Since emotions are instantaneous and automatic (for very good evolutionary reasons), they also tend to be closed; they sometimes foreclose potential routes of information acquisition. Emotion always assumes the amount of knowledge in hand is adequate to govern a decision, even when it may not be. Implicit in a lot of the new neuroscience is the message that we need to recognize moments when emotion is giving us a bum steer. And, like Odysseus, we need to think a few steps ahead so that we can push back against an otherwise invincible neural force.

Since so much of our decision-making apparatus is subterranean, this may sound like an impossible act of wisdom physics: pushing back against something that is not palpably there. But as we have seen repeatedly in other settings—cognitive behavioral therapy (which addresses habit), meditation (which builds up the "muscle" of cognition and empathy), compassion (which facilitates reframing), and even "thinking like a trader" (which trains the mind to overcome its emotion-driven aversion to risk)—we have an opportunity to tamper productively with this system.

The one thing all these interventions have in common, however, is effort: to attend therapy, to meditate, to enlarge compassion, to change habit and outlook. Surely wisdom, of all strains of human excellence, does not come easily and without effort. But neuroscience is now beginning, just barely, to identify areas where such effort can make a difference. As Cohen writes, "People do have the capacity to override emotional responses." "Capacity" is not birthright or entitlement; it is the possibility, embedded in the neural circuitry just coming to light, to make ourselves better through effort.

The idea of meta-wisdom is really an invitation to reframe, to step back and reassess a vexing situation from top to bottom. This process probably involves a reconsideration of our working knowledge and a commitment to acquire new information, however unsettling it might be, which thus implicates courage; it obliges us to mine different shafts of memory; it forces us to reevaluate all this intelligence and fact-finding and feeling in the face of uncertainty; and it encourages us to acknowledge the fact that context, the frame around the problem, is crucial. In short, meta-wisdom is an invitation, an imperative, to *deliberate*. And this, frankly, is where much contemporary neuroscience falls short, where it runs up against the mystery of wisdom.

Hardly any current research in social neuroscience and neuroeconomics addresses the notion of deliberation. At that three-day meeting of neuroeconomists at New York University in 2008, the only scientist I heard utter the word *deliberation* in the context of decision making in a talk was neuroscientist Antonio Damasio. In the vast majority of fMRI experiments that investigate decision making, subjects are asked to render their judgments in a matter of seconds. In any human dilemma worthy of the name, and worthy of wisdom, old knowledge or experience almost certainly needs to be reevaluated in light of a new situation. Our default neural setting is tradition and habit; our more adaptive responses probably lie in breaks from tradition and habit. Getting to that adaptive response requires new information, cultural clues, psychological flexibility, and, yes, deliberation. And finally, we must be open to change. We need to be willing to break habits—either habits of action or, more difficult, barely conscious habits of emotional thought. Uncertainty, if nothing else, can cure us of habit.

Montaigne was so enamored of Pliny's remark about uncertainty—"There is nothing certain but uncertainty, and nothing more miserable and arrogant than man"—that he had the sentence inscribed on the ceiling of his library. Our misery and arrogance derive in no small part from the tyranny of habit, and this inflexibility is at the heart of a lot of wrongheaded decision making. Jonathan Cohen's main point is that one apparatus of decision making (the emotional) is rapid but inflexible; Garry Kasparov is so locked into the rules and algorithms of chess that he is brilliantly fast, yet perhaps also imprisoned by inflexibility. The other apparatus, more cortical and top-down, is less rapid (deliberation and thought usually are) but also more supple. Wisdom is not only found in both these approaches; wisdom, Cohen suggests, also resides between them, at that ubiquitous fork in the road where we need to figure out which system is more appropriate to which situation, which path is likelier to lead where we ultimately want to go.

None of us can foretell the future, but moments of crisis and uncertainty always force us to review our portfolio of life decisions. When U.S. financial markets began to implode in the spring of 2008, employees of a prominent Wall Street investment bank lost their lifetime savings—for some, millions of dollars, for others more modest but zealously protected nest eggs of $400,000 or $600,000—in a matter of twenty-four hours. Such a swift elevator drop in net worth would put a heartsick wobble in anyone's knees, but I was particularly struck by a story in *The Wall Street Journal* that quoted one devastated broker as lamenting, "I've spent more time at Bear Stearns than I have with my own family."

Hindsight is easy, but any decision we make—in the neurological as well as the wisdom sense—is only as good as what we ultimately value, what we understand to be most important. It's hard to imagine how the brain finds a common scale to measure the relative value of time spent with loved ones versus that devoted to building a secure retirement account, and yet we are forced to make these judgments almost every day. It's a reminder that, although *Homo economicus* insists by definition on a narrow and material definition of "preference," *Homo sapiens* ultimately juggles a much more complicated set of values.

There is a simple answer to this dilemma in the Bible, where Wisdom herself reminds us, "Happy are those who find wisdom, and those who get understanding, for her income is better than silver, and her revenue better than gold."

But there is a more sobering assessment in the gospel of capitalism, where the young Adam Smith, in writing an epitaph for his age, essentially wrote one for ours, too. In a lovely evocation of that timeless fork in the road between material and spiritual well-being, he spoke of two different roads—one of "proud ambition and ostentatious avidity," the other of "humble modesty and equitable justice"—that await our choice. "Two different models, two different pictures, are held out to us," he continued, "according to which we may fashion our own character and behaviour; the one more gaudy and glittering in its colouring; the other more correct and exquisitely beautiful in its outline: the one forcing itself upon the notice of every wandering eye; the other attracting the attention of scarce any body but the most studious and careful observer. They are the wise and the virtuous chiefly, a select, though, I am afraid, but a small party, who are the real and steady admirers of wisdom and virtue. The great mob of mankind are the admirers and worshippers, and, what may seem more extraordinary, most frequently the disinterested admirers and worshippers, of wealth and greatness."

PART THREE

BECOMING WISE

Even if we could be learned with other men's learning, at least wise we cannot be except by our own wisdom.

—Montaigne

CHAPTER TWELVE

YOUTH, ADVERSITY, AND RESILIENCE

The Seeds of Wisdom

Animals whose hoofs are hardened on rough ground can travel any road.

—Seneca

THE BIOLOGIST MARIO CAPECCHI did not publicly disclose the details of his extraordinary childhood at any length prior to 1996, when the Italian-born scientist traveled from his lab at the University of Utah to Japan to accept the Kyoto Prize, a precursor to the Nobel Prize he would receive eleven years later. In an autobiographical lecture entitled "The Making of a Scientist," Capecchi stunned his audience by describing his upbringing in Europe in the 1940s, during which his single mother was arrested and dragged off to a Nazi concentration camp, his supposed guardians turned him out when he was four years old, and he lived for years as a homeless, malnourished, and itinerant street urchin in war-torn Italy—a story, he told his Japanese hosts, that "exemplifies the antithesis of a nurturing environment." In an unexpected way, his story also helps us think about the roots of wisdom.

Capecchi's mother, Lucy Ramberg, grew up in a villa in Florence, Italy, a child of privilege who went on to become a free-spirited poet. Her parents (Mario's maternal grandparents) were Lucy Dodd, an Oregon-born Postimpressionist painter who had expatriated to Italy, and Walter Ramberg, a German archaeologist of some repute who died during World War I. After growing up in Villa Ramberg, with its sprawling gardens and armies of nannies and servants and private tutors, Capecchi's mother left home for Paris, where she studied literature and languages at the Sorbonne, lectured at the university, and began to publish poetry (in German) in the 1930s as part of a politically active, stridently antifascist literary group known as "the Bohemians."

In 1937, Lucy Ramberg relocated to the small village of Wolfgrübben, north of Bolzano in the Italian Alps. On October 6, 1937, Mario was born in nearby Verona. His father, an officer in the Italian air force, had a brief affair with his mother and then effectively disappeared from her life and his. Two years after Mario's birth, he was followed by (it became clear only recently) a half sister named Marlene. As Lucy's antifascist activities courted greater and greater danger, she took steps to provide for an uncertain future by arranging for Marlene to be adopted by an Austrian family and for Mario to be raised by a nearby farm family if anything should happen to her. That day of foreseen reckoning arrived in the spring of 1941, when "the Gestapo and SS officers came to our home and arrested my mother," Capecchi revealed in his Kyoto speech. "This is one of my earliest memories. Although I was only 3½ [years old], I sensed that I would not see her for many years." The Germans sent Lucy to Dachau, he said, where she was held as a political prisoner, and Mario lived with the peasant family. At first, he enjoyed this rustic lifestyle, watching his adoptive family as they harvested wheat and happily participating in the autumn grape harvest. ("The children, including me, stripped, jumped into the vats and became squealing masses of purple energy," he recalled.)

His mother had sold many of her possessions to raise money to provide for Mario's care, but one year later, in 1942, the peasant family decided Mario had to fend for himself. "For reasons that have never been clear to me," Capecchi said in Japan, "my mother's money ran out after one year and at age 4½ I set off on my own." Capecchi said he lived the life of a homeless orphan for more than four years. "I headed south," he recalled, "sometimes living in the streets, sometimes joining gangs of other homeless children, sometimes living in orphanages, and most of the time being hungry. My recollections of those four years are vivid but not continuous, rather like a series of snapshots. Some of them are brutal beyond description, others more palatable." (If this story sounds improbable, it's worth recalling that Italy in the early 1940s was a scene of breathtaking poverty and wartime deprivation; Elsa Morante's *His-*

tory: A Novel provides a numbingly stark portrayal of the impact of World War II on Italian domestic life.)

In the spring of 1945, American troops liberated Dachau, and Lucy Ramberg, traumatized by years of incarceration, began a methodical eighteen-month search for Mario throughout Italy. She eventually found him in October 1946 in the city of Reggio Emilia (some 160 miles south of Bolzano), in a hospital filled with hundreds of listless, orphaned children; he had been hospitalized for malnutrition, racked by fevers, and had slept naked in a filthy bed for the better part of a year. She had aged so much in five years that Mario did not even recognize his mother when she first appeared at his bedside.

After making their way to Rome, where Mario had his "first bath in six years," mother and son booked passage on a ship from Naples to New York. Lucy Ramberg never recovered mentally from the experience; "she lived essentially in a world of imagination," as her son put it, and died in Arizona in 1989. Mario, meanwhile, was raised by his aunt and uncle, who lived in a Quaker commune outside Philadelphia; he later attended Antioch College, ultimately trained as a molecular biologist in James D. Watson's laboratory at Harvard University, and did pioneering work in gene manipulation and stem cells, which resulted in his winning the 2007 Nobel Prize.

Now in his seventies, Mario Capecchi is a small, shy man who, though not without ego, would be loath to describe himself as wise. Yet bits and pieces of this shattered biography shine with qualities we associate with wisdom: his deep and lifelong commitment to community service and social welfare, a social consciousness that was instilled in him during his Quaker education; his phenomenal ability (remarked upon by many colleagues) to focus on a single issue with Jamesian intensity, undistracted by anything else; his comfortable and confident sense of patience, which has allowed him to tackle important problems (scientific or otherwise) that usually don't lend themselves to speedy solutions; and, of course, his emotional steadiness and resilience in tackling these difficult problems. And so it is reasonable to wonder if his experiences early in life somehow primed him to develop the characteristics he has exhibited at an exemplary level as an adult. Capecchi himself has wondered about this.

"Looking back, I marvel at the resilience of the child," he said in

1996. "As I lay naked on that stripped bed so many years ago, my constant preoccupation was fixed on plotting an escape. In the absence of any apparent hope, the will to survive persists.

"It is not clear," he continued, "whether those early childhood experiences contributed to whatever successes I have enjoyed or whether those achievements were attained in spite of those experiences. When dealing with human life, we cannot do the appropriate controls. Could such experiences have contributed to psychological factors such as self-reliance, self-confidence or ingenuity? I have always considered as a personal strength the ability to concentrate for long periods of time on a chosen topic at the exclusion of everything else that is going on around me. . . . What I have learned from my own experiences is that the genetic and environmental factors that contribute to such talents as creativity are too complex for us to currently predict. In the absence of such wisdom, our only course is to provide all of our children with ample opportunity to pursue their passions and their dreams."

The example of Mario Capecchi's life—its horrendous beginning, its spectacular later flowering—challenges one of the more comforting pop-culture mythologies about wisdom: that it comes more easily with old age. It may indeed *come* more easily at a later stage in life, but there are both anecdotal and scientific hints that its genesis probably occurs, at least in some cases, much earlier, and is often associated with some sort of adversity. As Capecchi suggests, we can't ethically do "adversity experiments" on young children. But when we comb both the scientific literature and the literature of wisdom for clues about the seeds of wisdom, we are inspired to ask some more provocative questions.

What if those seeds are, in fact, planted early in life? Is part of the "older but wiser" formula "younger but resilient"?

Aristotle had a speech impediment and was orphaned at an early age; Moses stuttered. Socrates was famously ugly. Pericles had a head so narrow and congenitally misshapen that Plutarch describeed it as a "deformity" and recounted that the comic poets of Athens took malicious glee in calling him "*schinocephalos,*" or "squill-head." Gandhi lamented his frail boyhood body and a shyness so profound that other children laughed at his reticence. Confucius's father died when he was three, and Abraham Lincoln's mother died when he was nine. Siddhartha Gautama's mother died when he was seven days old, and even as a young adult, the future Buddha was virtually imprisoned by his own father, who was alarmed by a prophecy that his son would abandon both family and wealth in search of spiritual awakening. Paul Strathern, the author of a brief, brisk biography of Confucius, notes, "Curiously, of the dozen or so figures who founded the world's great philosophies and religions, a large majority were brought up in single-parent families." (Single parenting doesn't necessarily equate with adversity, of course, but, as any parent will tell you, it adds a certain degree of stress to domestic economy!) These are purely anecdotal scraps of biography, but they allow us to connect some dots between historical figures of widely accepted wisdom and childhoods of at least nominal adversity.

A number of psychologists who have dared to tackle the subject have been struck by the connection between wisdom and adversity. The resulting resilience, this ability to persist in the face of emotional challenge, this capacity for coping with the inevitable upsets and sorrows that life throws at us (with the rhythmic casualness of a batting-practice pitching machine) emerged in the 1990s as one of the central traits, for example, in Monika Ardelt's psychological measures of wisdom, and it has also captured the attention of experimentalists exploring the psychology—and neurobiology—of emotional resilience.

The Berlin Wisdom Project was the first (and probably only) modern attempt by a group of researchers to explore the origins of wisdom in any systematic way. Beginning in 1984, Paul Baltes and his colleagues at the Max Planck Institute for Human Development tried to identify common features in the backgrounds of people who had scored above-average ratings on the Berlin team's various wisdom measurements. Two surprising insights emerged from this research.

First, the seeds of wisdom appeared to be planted much earlier in life than during mature adulthood. In no fewer than four different studies that grew out of a concurrent research project, the Berlin Aging Study, for example, the Baltes group concluded that the kind of knowledge and judgment typical of wisdom appears in *early* adulthood and does not measurably increase over time. In fact, some research has located the roots of wisdom as early as adolescence or early adulthood. As Baltes and his longtime collaborator Ursula M. Staudinger reported in 2000, their research on adolescents "has suggested that the major period of acquisition of wisdom-related knowledge and judgment before early adulthood is the age range from about 15 to 25 years." Contrary to conventional (and cultural) expectations, old age does not automatically confer wisdom, and being young does not necessarily preclude it.

Second, wisdom often grew out of an exposure to adversity early in life. Many participants in the Berlin Aging Study who rated high on wisdom testing, according to coauthor Jacqui Smith, had lived through some of the twentieth century's most tumultuous events as children and young adults: the humiliating aftermath of World War I for Germans, the frenetic excesses of the Weimar Republic, the associated financial depression of the 1930s, the rise of Nazism, and the utter devastation of Germany's physical, social, and spiritual infrastructure by the end of World War II. "In the Berlin Aging Study," Smith told me, "we have some people who have lived through two world wars. *Two world wars!*" The data are too skimpy to support generalizations regarding different, subsequent populations, but Smith wondered aloud in a conversation whether people born in the post–World War II generation experienced enough adversity, or learned enough from the adversity of their parents' generation, to gain emotional resilience. "The baby boomers have life pretty good, right?" mused Smith, who worked on the Berlin Wisdom Project for many years and is now on the faculty of the University of Michigan. "We don't really know whether having life good is going to be actually good for us. Is a good life good for you? We don't know. Those things will play out, and somebody in fifty years' time or one hundred years' time may say, 'Ha-ha, the baby boom generation.' Perhaps. We don't know. These are people who have mostly succeeded. But what happens when you don't succeed?"

In his valedictory work on wisdom, Baltes attributed the acquisition of wisdom to a variety of factors—general intelligence and education, early exposure to meaningful mentors, cultural influences, and the life-long accumulation of experience, which is the centerpiece of developmental psychology. But he, too, acknowledged the central importance of emotional intelligence, noting that "there is good reason to assume that people capable of effectively regulating emotional states associated with dilemmas of life by cognitive rather than affective-dysfunctional modes might have a better chance of being considered wise or scoring high on wisdom tasks." That, in neurological shorthand, hints at the role of the prefrontal cortex in managing, steering, or otherwise herding the impulses of emotion.

Many elderly people who scored high on Monika Ardelt's three-dimensional wisdom scale also reported considerable hardship earlier in their lives. In one paper, she recounts the history of a pseudonymous "James," who, in addition to the social adversity of growing up as an African American in the South, was deeply traumatized by witnessing acts of violence and brutality during World War II service in the Pacific and suffered severe episodes of depression upon his return; he emerges in her study as an accomplished educator, successful athlete, and an adult with an inner calm and emotional serenity. Similarly, the experimental subject named "Claire" described a childhood of economic hardship on a tobacco farm in Kentucky during the Great Depression.

"Adversity early in life is an apprenticeship for old age," said Ardelt. In a study (conducted with George Vaillant) of Harvard University students who graduated in the 1940s, she said, "There seems to be a seed of wisdom in people. The people who were wise in mid-life and old age, they were already other-centered in their youth. If you have that seed of wisdom, it's easier for wisdom to bloom. But it doesn't happen automatically."

The traumatic biographies of ancient philosophers may tell us more about the generic difficulties of life in premodern times, when no life was immune to adversity, than about the roots of wisdom. And yet lessons in adversity permeate more recent psychological research as well, and aphorisms and maxims referring to adversity form part of the wisdom literature. In the view of the Roman philosopher Seneca, who is something of a bard of adversity, "It is a rough road that leads to the height of greatness."

But can the effects of adversity be tracked experimentally and described biologically? Mario Capecchi himself has confessed to "a deep-rooted prejudice" on the subject. He believes that complex human traits like creativity (and, presumably, wisdom) are too complicated to allow people to think they might be able to orchestrate or manipulate their experiences to cultivate these higher-order qualities.

Let's grant, temporarily, that he's right about humans. How about monkeys?

. . .

In 2001, five years after Capecchi speculated about the biology of human resilience, researchers at Stanford University began an ambitious long-term experiment to discover the origins of resilience in a group of squirrel monkeys. They arranged for twenty pairs of the primates to mate, cared for the females during the five and a half months of gestation, and then deliberately manipulated the twenty offspring in an effort to detect the long-term behavioral and biological effects of moderate early-life stress exposure. This kind of research is not only important but indispensable, because ethical concerns and practical limitations obviously restrict opportunities to study stress exposure in human children. However, it is precisely these kinds of studies that may provide clues about the way our neural plumbing processes adversity at various stages in our physical and psychological development. In fact, these scientists gave about half of the young monkeys a "stress inoculation"—like a vaccination for emotional sensitivity—to see what kind of developmental effects it would have.

Here's how the experiment worked: The Stanford researchers, led by Karen J. Parker, allowed the twenty infant squirrel monkeys, born in the Stanford animal facility, and their respective mothers to live undisturbed for seventeen weeks (in nature, these monkeys are able to move around independently by five weeks, forage on their own by seven weeks, and are weaned by week sixteen).

Then, at week seventeen, the scientists challenged about half of the young monkeys with repeated doses of moderate stress. Once a week for ten weeks, researchers separated the young monkeys from their mothers and placed them in a cage in another room adjacent to unfamiliar adults for an hour. This "intermittent stress inoculation," as the Stanford team called it, provoked predictable consternation: The separated young monkeys issued isolation calls, moved around in an agitated fashion, and released greater amounts of cortisol, the so-called stress hormone. Meanwhile, the control group of ten mother and infant monkey pairs remained unstressed in their "home" cages.

Several months later, Parker and her colleagues deliberately exposed *both* groups of young monkeys to a potent emotional challenge: an unfamiliar environment. In one experiment, performed roughly nine months after birth of the offspring, the researchers relocated each mother and child pair to a new cage in another room that contained both familiar and unfamiliar objects; Parker recorded the behavior of the offspring (how long they clung to their mother's back, how long it took them to investigate the novel objects, and so on) and took measurements of cortisol and ACTH, another stress hormone. In a second experiment, at week fifty, researchers placed an unfamiliar stainless-steel wire-mesh box next to the home cage of the monkeys and provided an interconnecting opening; Parker again observed how long it took for the now year-old monkeys to enter the new cage, how long they stayed, and how much they played.

The researchers didn't find any philosophers among the stress-inoculated monkeys, but they *were* behaviorally different. By the end of this elaborate yearlong experiment, the Stanford team found that the monkeys that had been briefly exposed to a stressful situation ("inoculated") about four months after birth handled anxiety much better when they encountered unfamiliar or uncertain situations later on. At the behavioral level, they were less clingy, more curious and exploratory, and devoured more cantaloupe and marshmallows (greater food consumption is a sign of low anxiety); at the physiological level, analysis of their blood showed lower levels of stress hormones in it than in the blood of the control group, so the behavioral differences corresponded with biological differences. "These are all indications of diminished anxiety," said Parker in an interview. "The stress-inoculated monkeys have better anxiety response inhibition." Indeed, all the monkeys who had been previously exposed to stress proved more intrepid, venturing into and playing in the wire-mesh box; none of the "unvaccinated" monkeys did so.

As the scientists pointed out, it's not that the inoculated monkeys were immune to stress—*all* the animals responded with anxiety and higher stress-hormone responses (and appropriately so) when confronted with a novel and potentially upsetting situation. It's that the monkeys who'd been inoculated against stress adjusted to novelty better and were able to "ride the horse" of emotional arousal more successfully. As Parker and her colleagues later put it, "These findings suggest that prior experience with moderately stressful events early in life results in more effective arousal regulation when emotionally challenging circumstances are again encountered."

. . .

You might reasonably wonder at this point if monkeys noshing on marshmallows in a wire-mesh cage really tell us anything about human wisdom. You will recall that emotional regulation—the ability to modulate emotional responses and stay even-keeled in the face of conflict or stress—has emerged as a key coping mechanism in older adults. That's not to say that this capacity is inevitably and only shaped by early-life events; it does suggest, however, that the emotional circuitry of the primate brain can be reengineered by early experience, so that this aspect of wisdom can, in some cases, legitimately claim very early roots.

Moreover, follow-up research by Parker and her colleagues at Stanford (with the same set of monkeys) suggests that this stress-induced coping mechanism is in part a *cognitive* trait. In other words, the prefrontal part of the stress-inoculated monkeys' brains seemed to be exerting some sort of moderating influence on the stress reaction. They teased this fact out by challenging their monkeys with an experimental task known to require prefrontal cortex activity to solve a potentially frustrating problem—the appearance of food inside a clear plastic box, which the monkeys could obtain only by reaching in from the side. Impulsive monkeys lunged straight for the food, found their hands blocked by the clear side of the box, and often grew frustrated by their inability to get to the food. "Our stress-inoculated monkeys had a better ability to control their impulsive response," Parker said, "and just anecdotally, these guys got less frustrated." Unpublished data suggest that this form of cognitive control lasts at least three years, and preliminary brain scans showed "a very selective enlargement" of the ventromedial prefrontal cortex in stress-inoculated monkeys—a preliminary but dramatic suggestion that early-life experiences can produce measurable changes in brain physiology. "If it happens earlier in life," Parker said, "it has a better capacity for programming your brain."

As a general trend, other research is beginning to fill in the picture of how these early exposures to stress shape long-term biological effects, both positive and negative. Bruce McEwen of Rockefeller University and colleagues at the University of New Mexico did experiments several years ago showing that exposing rats to a novel situation shortly after birth produced dramatic, long-lasting neurological changes, particularly an increased capacity for social memory, which persisted into adulthood. In these animals, brain activity changed when they encountered a new environment, and there were measurable alterations in the hypothalamic-pituitary-adrenal (HPA) axis, which douses animals with hormones like cortisol and adrenaline when they are stressed. But early stress can cut both ways. Researchers at the State University of New York Downstate recently reported that exposing very young monkeys to food stress early in life primed the animals to develop obesity and insulin resistance at the time of puberty.

This entire school of animal research has grown out of what might be called "the puzzle of human resilience." Beginning around 1970, psychologists began to document instances in which an earlier exposure to stress diminished later stress responses—less anxiety in children and adolescents who had been admitted to a hospital, less emotional response to stressful laboratory tests, less post-traumatic stress in people who had been torture victims, even less anxiety among flood and earthquake survivors who had previously experienced those natural disasters. Some of these stress-related changes can leave a mark on developing young brains.

Nearly thirty years ago, Richard Davidson (whose work was discussed in the chapter on compassion) identified a difference—an "asymmetry," as he put it—in electrical activity between the left and right sides of the prefrontal cortex. Greater left-sided activity correlated with a more upbeat, positive emotional tone, and this difference could be discerned as early as in infancy. In a fascinating study published back in 1989, Davidson and Nathan A. Fox tested emotional resilience in ten-month-old babies by separating the infants from their mothers and then closely observing their subsequent behavior. They discovered that the infants who coped best with separation—who didn't cry and simply began to explore their new environment following separation—were the same ones who had displayed greater left-sided prefrontal activity and less right-sided activity during an earlier experimental task. By contrast, the infants who bawled inconsolably when separated from their mothers— who displayed, in other words, less resilience—showed the opposite pattern of brain activity.

This isn't to suggest, I hasten to add, that our quotient of emotional resilience is set in stone even while we're still a mound of baby fat. As

Davidson's own research shows, these distinct patterns of brain activity, while generally stable in adults, bounce all over the place in growing children. In a study that lasted eight years, Davidson and his colleagues measured this kind of brain activity in children—the same children—from the time they were three-year-old toddlers until they reached the cusp of adolescence at age eleven. The bottom line was that "little evidence of stability was found." These findings suggest that the neural circuitry related to resilience and other aspects of emotional behavior, which is known to involve conversations lasting milliseconds between the prefrontal cortex, the amygdala, and the hippocampus, is exceedingly "plastic" (or variable) during childhood and adolescence. "This is a period," Davidson has written, "during which pronounced plasticity is likely to occur in the central circuitry of emotion, particularly in the PFC [prefrontal cortex], which is still undergoing important developmental change at least until puberty." If the prefrontal cortex rides the horse of emotion, it seems like the riding lessons we accidentally and even involuntarily learn during adolescence are particularly crucial. But it's also clear that we're nowhere near any definitive answers about this. "One of the major challenges for human affective neuroscience in the next century," Davidson observed in 1995, "is to better understand the environmental forces that shape the circuitry of emotion."

The Stanford research on squirrel monkeys appears to be the first attempt to explore the biology of stress inoculation in a designed, empirical, and prospective (that is, not after the fact) experiment. But before we get too excited about a few studies based on your basic barrelful of monkeys, a few caveats. A depressingly deep psychological literature makes clear that early stresses in human life can trigger a textbook's worth of human pathology; resilience is a puzzle precisely because some people gain strength from these early challenges, while others seem permanently harmed by them.

There is also good, but possibly conflicting, evidence that the quality of maternal care during early postnatal development can be an important buffer against early stress, too. A beautiful set of animal experiments by Michael Meaney's group at McGill University, in Canada, have shown, for example, that maternal nurture, where mother rats lick their pups, actually induces permanent biological changes in newborns; the licking alters the pattern of gene expression in the HPA axis of the infant animals (this is the same stress-coping circuitry explored in the Stanford monkey experiments). It's tempting to say, only partly tongue in cheek, that every good mother is a source of wisdom, but the Stanford experiments seem to argue that the benefits of stress inoculation, at least in primates, may be independent of maternal care.

Finally, as you might expect in any discussion of wisdom, context is critical, and the confounding variables are enormous. As Parker and her colleagues concede, the "type, timing, duration, and severity of a given stressor within a given species are likely to be important factors in determining whether early experiences ultimately produce a protective or deleterious outcome."

No matter how early wisdom *can* form, however, that doesn't mean it necessarily *does* take shape early in life (if, of course, it takes shape at all). Life-span psychology—the way our emotional core evolves over the course of a lifetime—is particularly instructive here. As Laura Carstensen's elegant longitudinal studies of "emotional regulation" demonstrate, older people tend to be more emotionally evenhanded than their younger counterparts, in part because of the way their shrinking horizon of time (and the increasing imposition of emotional loss) nudges them toward more positive and meaningful emotional interactions. And as Fredda Blanchard-Fields's work has repeatedly demonstrated, older people deploy this emotional strength to resolve conflict and solve everyday relational or social problems in a way that younger people do not. If we assume that the same fundamental emotional circuitry is in play in both adults and children (an assumption that will likely require tinkering and troubleshooting as we learn more about the brain), it seems plausible that resilience, coping, or indeed any form of emotional regulation might ultimately be *learned.*

If we think about resilience from an evolutionary point of view, we are struck by a fundamental truth that is reflected incidentally in many of the philosophical aphorisms about adversity, from Horace to Poor Richard. Avoiding stress is probably not an issue that natural selection cares about, because stress is unavoidable in virtually any dynamic environmental situation where new circumstances and threats constantly arise. Some theorists argue, however, that natural selection might care about cultivating a neural mechanism that could modulate and, in a sense, master the emotional experience of risk. Whether that mastery is partly acquired early in life, as the stress-inoculation research suggests, or later in life, as research by Carol Ryff and her colleagues at the University of Wisconsin has found, the end result is an enhanced form of emotional regulation that would clearly confer adaptive power on anyone who possesses it.

We're a long, long way from prescriptive remedies like "The Ten-Week Wisdom Curriculum! Do it at home in your spare time!" But lurking just beneath the surface of this research—as well as beneath the work on compassion, attention, and delayed gratification—is the notion that some sort of practice, some kind of proactive and systematic mental training, has the potential to change the way our minds handle the challenges of adversity. Some might claim those practices are already in place; I think it's too early to reach any conclusions. But the excitement of this kind of research lies precisely in the fact that we are beginning to sense a *biological* rationale for mental practices that might enhance our emotional regulation and our cognition. Someday, in a still distant future, it might even help us become wiser.

The suggestion that adversity early in life can vaccinate us from later stresses and help us become more emotionally resilient is a powerfully counterintuitive observation, one that may force us to rethink ideas about how wisdom develops, both in the abstract and in our own lives. And the provocative idea that wisdom might even be cultivated begins to change the cultural perception of the entire topic. No matter how old we are, no matter what life circumstances we find ourselves in, all of us, including children, are at least theoretically in the process of becoming wise. As Heraclitus observed, "Time is a game played beautifully by children," and perhaps a child's early exposure to adversity somehow establishes the beat with which he or she taps into wisdom.

On the other hand, we should not become too enamored of precise and determinative biological explanations for something as elusive as the origins of wisdom. In Book I of the *Meditations,* Marcus Aurelius laid out the beautiful genealogy of his personal values, which the worlds of both literature and philosophy have rightly judged to be exceptionally wise: from his grandfather, "the lessons of noble character and even temper"; from his father, "modesty and manliness"; from his mother, "piety and bountifulness." His serial gratitude goes on for several pages, and subtly makes the point that whatever wisdom we manage to achieve derives from genes, nurture, mentorship, culture, and, perhaps most of all, an openness to the possibility of continual learning and self-improvement.

Despite that rich lineage, there is a sobering paradox about the wisdom of Marcus Aurelius that speaks not so much to its origins as to its limited shelf life. Gibbon famously wrote that in all of history "the condition of the human race was most happy and prosperous" during the reigns of the two Antonines, Antoninus Pius and his adopted son, Marcus Aurelius. Yet, perched alone atop that monument to human happiness, Marcus Aurelius never shed the cowl of melancholy that enveloped him like a monk's hood and that permanently colors our view of him. And for a man whose *Meditations* survive as an unusually deep reserve of wisdom, Marcus Aurelius, as historians have been forced to concede, could show terribly unwise judgment. In arguably the two most important political appointments made during his time in public office, he designated Verus, a politician hopelessly over his head, to be coemperor, and, even more disastrously (as any viewer of the film *Gladiator* will recall), designated as successor his malevolent son Commodus, whose jealousy and paranoia turned Rome into an imperial bloodbath. As Gibbon remarked, "The monstrous vices of the son have cast a shade on the purity of the father's virtues."

The connection between the childhood of Mario Capecchi and the biology of human resilience is, I admit, as thin as anecdotal spittle. But there is a coda to his story that is very revealing about Capecchi's own powers of emotional regulation. Following the announcement of his Nobel Prize in the fall of 2007 (and the numerous press stories that ensued about his unusual childhood), journalists investigated the veracity of his childhood memories and uncovered "several inconsistencies." The Associated Press reported that on several occasions during his "homeless" period, Capecchi, in fact, stayed briefly with his father (he remembers it as "three times, each for a duration of about at most a week"); that there is no record that his mother was a prisoner at Dachau (although there is little dispute that she was arrested and probably deported to Germany); and that it couldn't have been the Gestapo who arrested her, because that arm of the German police wasn't operating in Italy in 1941 (although common sense suggests that the overwhelmingly salient emotional fact here, especially to a three-year-old child, was that hostile soldiers came and dragged his mother away, not which branch of the Nazi war machine the soldiers belonged to).

What is fascinating, however, was Capecchi's reaction when a reporter from the Associated Press visited his Utah home to confront him with this new information. Far from the cliché of the entrapped public official, shouting nondenials from behind a closed door, Capecchi invited the reporter in. During a lengthy interview, the AP reported, "Capecchi took an almost scholarly interest in the AP's findings, poring patiently over dates and matching his recollection against the historical record. At no point did he become defensive or uncooperative."

That may or may not be a reflection of wisdom, but it certainly sounds like someone who has a good grip on his emotions, and who perhaps learned that wise skill at a very early age.

CHAPTER THIRTEEN

OLDER *AND* WISER

The Wisdom of Aging

*We are all happier in many ways when we are
old than when we are young. The young sow wild
oats. The old grow sage.*

—Winston Churchill

*The advice of the old is like the winter sun:
it sheds light but does not warm us.*

—French proverb

LET US SAY you are a woman of a certain age, a grandmother, in fact, and one of your daughters-in-law has just given birth to her first child— your fifth grandchild! You go to the hospital and, while visiting your son and his newly enlarged family in the maternity ward, you proudly step forward to cradle your newest grandson when he is proffered. You have held babies all your life, including your own three children. You rock him. You hold him to your shoulder and pat his back. As he gets more comfortable in your arms, you gently jiggle and bounce him while cooing in his ear.

Suddenly, you hear the unmistakably chill voice of your daughter-in-law saying, "Let me have the baby, Mom. *Please!*" Beneath the perfunctory politesse, you hear a different emotional message: "Good grief, Mom, you're going to break his neck! Let me have the baby *now!*"

In the grand scheme of Socratic interrogatives and categorical imperatives, this little neonatal spat may not seem like a Yoda moment demanding exceptional wisdom. But you suddenly face a split-second decision, one that potentially could have long-term familial repercussions. Do you tell this daughter-in-law, this twenty-something maternal newbie, that you damn well know how to hold a baby, and have been changing diapers, thank you, since long before they were disposable and scented? Or do you simply pass the infant to his mother with the most grandmotherly of smiles, quietly leave the room—and then vent like a banshee at the other end of the hall, out of earshot, and wait until later to say something (discreetly, of course) to your daughter-in-law?

Fredda Blanchard-Fields has pondered precisely such situations—not as a chastised grandmother, mind you, but as a social psychologist who has incorporated the latest knowledge about the influence of emotion on social cognition into a series of clever experiments. Since the 1990s, her group at the Georgia Institute of Technology has conducted numerous studies comparing the way young adults and older adults respond to situations of stress. Blanchard-Fields cooks up scenarios of conflict typical of families and other social relationships, then probes how people of different ages respond to these everyday dilemmas. What she has found—with a consistency that should be heartening to any baby boomer—is that even though older adults suffer those well-known cognitive declines as they enter their sixties and seventies, they nonetheless seem to balance these physiological losses with measurable gains in social knowledge and emotional judgment, which actually increases their problem-solving skills. Adults between the ages of sixty and eighty, according to her research, think more strategically (and successfully) to solve problems, both impersonal intellectual problems and interpersonal social problems, than do young adults between the ages of eighteen and twenty-seven.

At one level, these results barely qualify as a latter-day footnote to Cicero's view that great deeds derive from thought, character, and judgment, and "far from diminishing, such qualities actually increase with age." Indeed, "older but wiser" has been a pat, self-flattering phrase in the psychological thesaurus of aging narcissists for centuries. But conventional wisdom and even literary bons mots often wither in the face of rigorous empirical testing, and few researchers have bothered to scrutinize the emotional intelligence of older people in as rigorous and empirical a manner as Blanchard-Fields.

Her results show that older adults are more socially astute than younger people when it comes to sizing up an emotionally conflicted situation. They are better able to make decisions that preserve an interpersonal relationship. Like Laura Carstensen, she has also found that older adults are able to behave in a more emotionally evenhanded manner than are young people. And she has found that as we grow older, we grow more emotionally supple—we are able to adjust to a changing situation on the basis of our emotional intelligence and prior experience, and therefore make better decisions (on average) than do young people. Older people are, in short, "more effective" problem solvers, in a way that very much aligns with psychological notions of wisdom.

Which brings us back to the grandmother in the hallway. As Blanchard-Fields noted, "The older woman's primary use of emotion regulation was effective (and perhaps wise) given the context and her goal to avoid a fight."

Wisdom may ultimately be a cognitive version of *Beat the Clock,* a pitched contest between life experience and mental decay, greater insight and shrinking cognitive ability, enlarged spirit in diminished body. In the dour view of Marcus Aurelius, "the apprehension of events and the ability to adapt ourselves to them begin to wane before the end."

Prior to the 1980s, the study of mental performance in the aged was a one-way street pointed toward pathology, with only one trajectory (downhill), one destination (senility), and one ending (unhappy). With the advent of life-span developmental psychology, which embraces the entire arc of an individual's life and views it within the person's social, cultural, and historical context, the picture became much more nuanced and complex. Pioneering psychologists like James Birren and Paul Baltes started looking for, and finding, mental and cognitive qualities that actually improve as we grow older. Indeed, modern empirical research on wisdom grew out of the realization that age conferred certain emotional and even cognitive advantages.

As Blanchard-Fields is the first to admit, a seventy-year-old brain is not the same as that of a twenty-something. It is slower to react. It is more likely to be distracted. But Blanchard-Fields has been among the first to investigate what others have merely speculated upon: that there's more to old age than cognitive decline, and that if you come up with a different kind of metric, you might even be able to identify some significant differences in the socioemotional performance of older adults.

The types of decline that come with aging are well-known to neuroscience. (Indeed, when scans of my own brain appeared in the pages of *The New York Times Magazine,* a physician friend cheerfully pointed out how the images confirmed two common features of the aging brain: its shrinking overall volume and its expanding ventricles, the fluid-filled cavities that surround the neural tissue. I was forty-seven years old at the time, but I felt like an octagenarian who'd just been told he'd dropped his drawers in public without even knowing it.) Sherwin B. Nuland cataloged these changes with depressing precision in *How We Die:* The brain loses 2 percent of its weight with each decade after age fifty, and cortical neurons in particular vanish like snow on a sunbathed sidewalk—20 to 50 percent of neurons in the motor cortex, about 50 percent in the visual cortex, about 50 percent in the physical sensory part of the cortex. "The end result is that the brain is smaller than it was in youth," Nuland wrote, "and doesn't work as well."

It's little consolation that anyone past the age of twenty is already well along the downward slope to forgetfulness and distractability. The passage of time subtracts from brain function in numerous and nefarious ways. With aging, the speed of mental processing declines. Several forms of neural attention, crucial to discerning what is important and how to adjust to multiple areas of focus, begin to flag. Several forms of memory show signs of faltering—not so much the recall of names and phone numbers, which seems to worry so many, but "working memory," which allows us to retrieve and process knowledge related to the task at hand. Sensory processing declines. Nuland held out modest hope about recent research suggesting that even in advanced age, the brain is able to generate new neurons in the cortical redoubts of greatest cognitive power; he even linked this to wisdom. "Neuroscientists may actually have discovered the source of the wisdom which we like to think we can accumulate with advancing age," he wrote in 1994.

Unfortunately, even more recent research is not quite so cheerful on this point. The part of the brain most vulnerable to the effects of aging happens to be the part most crucial to higher-level decision making, planning, resisting impulse, and exercising patience: the prefrontal cortex. Indeed, the dorsolateral prefrontal cortex, which some scientists view as a key player in many of the neural conflicts between the cognitive brain and the emotional brain, is believed to be the single most vulnerable segment of the prefrontal cortex to age-related decay. In other words, one of the biggest guns in the brain's armamentarium of wisdom, to hear some neuroscientists tell it, is among the first to show neural rust.

Even "gerontologically correct" psychologists have failed to find evidence of increasing wisdom with age. As part of the Berlin Wisdom Project, Paul Baltes and his colleagues conducted several studies exploring the relationship of wisdom to age, and they repeatedly failed to find any convincing evidence that wisdom, by their measures, increased much at all from the age of twenty to the age of ninety. It may be something we intuitively believe, but there's no empirical data to support it; as Baltes and his colleague Uta Staudinger put it, "having lived longer in itself is not sufficient for acquiring more knowledge and judgment capacity in the wisdom domain." Moreover, Baltes observed—and it is widely acknowledged—that, on average, cognitive performance begins "a more broadly based decline" around age seventy-five. So the potential window for wisdom begins inexorably to close for a lot of us about ten years after retirement.

Given this grim pile of scientific data, "older but wiser" seems like a paper-thin cliché, transparently untrue and one more example of a beautiful story, in the idiom of Peter Medawar, that just happens not to be true. But there is another part to this story, and the oft-maligned psychologists got there before the neuroscientists.

As far back as the 1980s, the Baltes group in Berlin conducted studies showing that, with some modest training, older people could improve their cognitive performance up to the level of *untrained* younger people (in other words, all things—including training—being equal, younger people, on average, will always outperform older people in certain cognitive tasks, but older people can significantly rejuvenate their cognitive chops through practice). More recent, and more sophisticated, studies have added nuance to this story. Jessica Andrews-Hanna and her colleagues at Harvard University recently documented a deterioration in communication between parts of the brain in older people, but they also reported that a surprisingly sizable portion of aged people in their sample bucked this trend and maintained good neural wiring. This combination of reasonably intact hardware and self-improvement software (in the form of cognitive exercise) helps explain why Michael Merzenich, a prominent (and recently retired) neuroscientist at the University of California at San Francisco, has argued that older adults who do mental exercises—what he calls "neurobics"—can improve their mental performance. Indeed, a cottage industry has grown up around training programs for older adults to maintain their cognitive tone.

These cognitive training programs, however, mask an even more fundamentally important message from the research community: Cognitive performance is not just about cognition. Emotion infiltrates and influences virtually every aspect of cognition, including memory, reasoning, and executive function. Blanchard-Fields and Carstensen have documented comparatively better performance among older people when it comes to devising strategies for solving problems, precisely because older people tend to process emotion differently.

Blanchard-Fields points out, for example, that recent neuroscience research has established a central role for the ventromedial prefrontal cortex, which coordinates a lot of emotional processing; studies have shown that this part of the brain remains relatively intact in older people, even as the dorsolateral PFC may be showing its rust. So while neurology textbooks still describe declines in "mental flexibility" among the aged, some fairly rigorous recent psychological research has found evidence of exactly the opposite: an enhanced ability (compared to that of younger people) to perceive contextual differences in day-to-day problems. Even the cultural vocabulary is beginning to change. The neuroscientist Elkhonon Goldberg has proposed the term *neuroerosion,* rather than *neurodegenerative,* as a more accurate way of capturing the decline of cognitive performance as we age—gradual rather than sudden, selective rather than global, often more of an annoyance than an outright impairment.

There is, however, a built-in limitation to the work of Blanchard-Fields and others. The research is like a snapshot of old and young people side by side, a transient X-ray of their social psychology and cognitive performance. In order to capture the psychological changes—including the kind of changes that might lead to wise behavior—within individuals as they go through life, we would need to conduct what scientists call "longitudinal" studies. These long-term experiments follow (and test) the same people over and over again as they pass from adolescence into middle age and sometimes into old age.

Longitudinal studies are rarely undertaken; they are complex, difficult, and enormously expensive. One such study, however, predates the era of brain-scanning technology, predates modern neuroscience, indeed predates the start of the baby boom generation. It was launched in Cambridge, Massachusetts, before World War II, and although it would be a stretch to say that it concerns wisdom, the study has generated a wealth of data about the kinds of behaviors and life choices that lead to a happy—or unhappy—life. It also, surprisingly, serves as an improbable bridge between Sigmund Freud's theory of defense mechanisms and modern wisdom research.

In the late 1930s, physicians at Harvard University began an ambitious longitudinal study that traced the occupational, cultural, physical, and psychological development of several hundred young men who embarked on the experiment when they were sophomores at Harvard, between 1939 and 1944 (one of the anonymous participants is said to have been John F. Kennedy). The Grant Study of Adult Development, as it was originally called, is still going on. Every two to four years (up to the present), the surviving men fill out questionnaires or submit to detailed interviews on everything from their career trajectory and cultural values to their satisfaction with family life.

In the 1970s, when the men had reached middle age, Harvard psychiatrist George E. Vaillant conducted in-depth interviews with those in the study to assess their various cognitive and emotional strategies. *Adaptation to Life*, Vaillant's book-length account of those privileged middle-aged men, is now barely known outside psychology circles, but I was intrigued when Vivian Clayton referred to a copy on her office bookshelf. It turns out to be a marvelous document, a kind of prosodic, Ivy League prequel to the famous British film documentary *Seven Up!* and its sequels. The book chronicles the decisions and the personalities of its subjects and then charts their resulting life trajectories. With its richness of detail and almost novelistic recounting of personal life histories, it essentially becomes a kind of psychological test that even viewers at home can score, and it has made an enormous impact on people who think about wisdom. George Ainslie was so taken with the power of the study that he once discussed a collaboration with Vaillant, and Clayton later told me, "There's a lot of wisdom in there."

The Grant Study researchers made no bones about its inherent biases—in Vaillant's words, it focused exclusively on "people who are well and do well." But despite its narrow sociological aperture, the main points to emerge from the study are very pertinent to the domain of wisdom. Life was unpredictable for everyone in the study, not surprisingly, but certain people in the Harvard cohort seemed to be able to adapt more successfully to uncertainty than did others. The successful adapters did so by using productive tricks; the less successful displayed maladaptive strategies that didn't do them, or their larger communities (families, coworkers, or social networks), much good. Vaillant's interpretation of these psychological strategies is heavily inflected by Freudian psychoanalysis; he characterized various emotional strategies as sublimation, passive aggression, dissociation, projection, fantasy, and other "ego mechanisms of defense." If you push past the equivalent of ancient Latin in the psychological vocabulary, however, you'll find a lot of behavioral meat on these pedigreed Harvard bones.

Emotional coping mechanisms, Vaillant pointed out, evolve and change over the course of a lifetime, even though people don't consciously "choose" them. He compiled a hierarchy of these strategies, from less to more successful, and associated some of the more sophisticated defense mechanisms with increased emotional maturity and more positive life fulfillment. Vaillant concluded that "a man's adaptive devices are as important in determining the course of his life as are his heredity, his upbringing, his social position, or his access to psychiatric health." In an unusually prescient remark for a Freudian psychiatrist in the 1970s, Vaillant even suggested that these healthier, mature adaptations went hand in hand with "an alteration of the user's nervous system." That insight anticipated the widely accepted notion now that learning and maturation are accompanied by—indeed, require—physical changes in the brain, a process known as "neural plasticity."

It's tempting to dismiss *Adaptation to Life* as hopelessly dated by its Freudian vocabulary and fatally clouded by its psychoanalytic lens. At the risk of seeming to dwell on an irrelevant cul-de-sac in the psychology of wisdom, however, I want to linger a bit on the Grant Study, because many of its findings, despite the archaic terminology, actually support and confirm many of the fundamental behaviors psychologists associate with wisdom, although it approaches them from an unexpected and, in the view of some, somewhat discredited direction.

As George Ainslie likes to point out, Freud remains valuable not for the answers he provided, but for the questions he asked about human behavior. He recognized, in what he called "defense mechanisms," psychological traits that we might now see as contributing to wisdom. He realized that these mechanisms are attempts to control instinct and emotion (that is, a way to try to rein in the elephant or grasshopper or horse—whichever is your preferred zoological metaphor); that they are often unconscious; and, most important, that these behaviors could be *adaptive* rather than being simply pathological, allowing a person to make healthy adjustments to a constantly changing and conflict-filled world. On the rare occasions when these adaptive mechanisms are conscious, deliberate efforts to outmaneuver our own instinctual nature, they harken back to Ainslie's cognitive reinterpretation of Ulysses and the Sirens: We know our own emotional inclinations so well that we have to concoct a strategy (presumably in the prefrontal cortex) to outsmart them.

It was Freud's daughter Anna who ultimately teased out the implications of these behaviors after her father's death. In the 1950s, she proposed that "denial, distortion, and projection were the defenses of psychosis, and at the opposite end of the continuum, sublimation, altruism, humor, and suppression were the defenses of maturity." The *possibility*—not inevitability—of a gradual, continuous evolution of these behavioral strategies as we age, from the immature tantrums of childhood and adolescence to the more mature forms of emotional regulation in middle age and beyond, represents one of the most powerful, albeit implicit, psychological arguments for the greater likelihood of finding wisdom in older adults, just as its stepwise progression does not preclude wisdom at a younger age in those who, by dint of circumstance or personality, clamber up the ladder of maturity more rapidly.

Still, how exactly does this Freudian notion of defense mechanisms intersect with more modern notions of wisdom? Vaillant, echoing Anna Freud, came around to the view that successfully mature adults displayed such emotional strategies as "altruism, humor, suppression, anticipation, and sublimation." Sound familiar? In the traditional literature of wisdom, we have encountered altruism countless times. Sublimation might just be a fancy, Germanic way of saying emotional regulation (Vaillant defined it as a skill where feelings were "channeled rather than dammed or diverted"). We could substitute other familiar terms—*patience, goal-directed thinking, other-centered behavior,* and *humility*—for some of the other defense mechanisms. In people who display such emotional strategies, Vaillant continued, "these mechanisms integrate the four sometimes conflicting governors of human behavior—conscience, reality, interpersonal relations, and instincts."

To Vaillant, these qualities were "convenient virtues." To our ears, however, they may sound like the by-now-familiar fundamentals of wisdom—conscience aligns with moral judgment and social justice, reality with discernment, interpersonal relations with empathy and emotional regulation, and instincts with the role of emotion in decision making. It seems to me that Vaillant and Freud are talking about the same thing (wisdom) in a slightly different professional dialect. True, the Grant Study men adopted their coping strategies unconsciously; still, the best strategies produced the most successful lives.

Perhaps the most fascinating aspect of the Grant Study is how the presence, or lack, of these "wise" defense mechanisms affected the lives of the Harvard men by the time they reached middle age. The men who exhibited "mature defenses," Vaillant reported in 1977, were happier, more satisfied with their careers and marriages, and "were far better equipped to work and love" than their peers who possessed less mature adaptations. They earned better incomes, engaged in greater public service, had more rewarding friendships, suffered fewer problems in terms of physical and mental health, and were even much more comfortable being aggressive with others, compared to men with less mature coping skills. One of the men, the pseudonymous "David Goodhart," for example, came from a modest blue-collar midwestern background, Vaillant wrote in *Adaptation to Life,* and yet displayed the more mature mechanisms of altruism, humor, and emotional coping, and was "unusually skilled in describing his feelings." Despite his "fearful, lonely" childhood, Goodhart went on to become a national leader in advocating public service for lower-income inner-city residents, and he consulted for the Ford Foundation. As Vaillant slyly observed of their first meeting, "despite his self-effacing qualities, Mr. Goodhart had kept me respectfully aware that I was talking to a man wiser and more experienced than myself."

As he struggled to summarize a mountainous heap of data and observation, Vaillant was struck by several overarching conclusions. Contrary to conventional wisdom, he found that "isolated traumatic events rarely mold individual lives." He found it useful, even as a psychiatrist, to think less in terms of mental illness and more in terms of emotional habits, which he called "characteristic reaction patterns to stress." And he acknowledged that adapting to challenge in some ways was the crucial determinant of a successful life. "It is effective adaptation to stress," he wrote, "that permits us to live." Evolutionary anthropologists, speaking of the survival of our species during prehistory, have made precisely the same point.

For all that, Vaillant remains at best agnostic on the question of whether we become wiser as we get older. In his 2003 book *Aging Well,* he followed up on the Harvard men and looked at data from two other long-running studies, including women from the Stanford-based Terman Study of gifted children. He acknowledged that later in life we cultivate broader social circles and manage to cope better with adversity. But after reminding readers that "wisdom involves the toleration of ambiguity and paradox," he concluded, "To be wise about wisdom we need to accept that wisdom does—and wisdom does not—increase with age." This hedging, enigmatic remark seems to suggest that wisdom is likelier, though hardly *likely,* as we grow older.

To knit together the Grant Study and modern psychological research on wisdom, the sociologist Monika Ardelt recently collaborated with Vaillant on several follow-up studies. In one study, their analysis of the Harvard men showed no correlation between religious practice and mental well-being. In a second, ongoing study, Ardelt is assessing how the men in the Grant Study rate in terms of charitable impulses (the quality she calls "other-centeredness" in her three-dimensional wisdom scale). Her preliminary analysis has turned up a strong correlation between those same mature defense mechanisms identified by Vaillant and a more charitable, compassionate pattern of behavior. This other-centeredness was independent of wealth, she found; some well-to-do Harvard men were especially effective in their charitable donations and activities, while others came from more modest backgrounds.

. . .

If wisdom does—and does not—increase with age, where does that leave those of us on the threshold of our sixties, waiting in line for that Golden Pass to Wisdom that would seem to be every baby boomer's birthright?

The research over the past twenty years on the importance of emotion in social cognition has pushed open a door previously bolted shut by our cultural obsession with pathology. By designing more sophisticated psychological studies, researchers like Fredda Blanchard-Fields and Laura Carstensen have repeatedly shown that there is, in fact, a difference between older and younger adults in their cognitive approach to solving problems—especially problems that are social and emotionally charged. These are precisely the kinds of situations that often demand (however loosely we define it) wise decisions—not in the Socratic sense of abstract ultimate truths, of course, but in the more modest, pragmatic Aristotelian sense of doing the right thing in order to live a good, meaningful life.

The recent psychological research has also shown that older adults feel more comfortable than younger adults do when dealing with uncertainty and ambiguity, surely one of the fundamental strengths of a wise temperament. It has shown that older adults are more supple in their assessments of problems; they are able to perceive the social context of a situation better than younger adults, and to adjust their actions accordingly. And perhaps most important, when it comes to settling on a strategy of action, they display greater flexibility, guided in part by their ability to regulate their emotions. In Blanchard-Fields's useful idiom, older adults develop a *feel*, something akin to emotional intuition, that lets them know when it is best "to do" and when it is best to "let it be." Which may be a softer, more psychological rendering of "Should I stay or should I go?"

It is tempting to dismiss this "let it be" philosophy of older adults as a senile version of conflict avoidance—a conclusion that seems even more warranted when, as in one of Blanchard-Fields's experimental scenarios, you find out that your spouse is having an affair and you simply decide to "let it be." In the unsentimental view of neuroscience, it is even possible that this emotional flexibility is a fortuitous side effect of mental decline during old age.

But it is possible to look at the not-uncommon crisis of spousal infidelity, as social psychologists have begun to do, and argue that "let it be" may be precisely the wisest course of action, depending on the context of the situation: age and station in life, the presence of children, financial considerations, indeed the whole panoply of present and future variables that make wise decision making so daunting. What this research argues is that the first step toward wisdom may indeed be the temporary suppression (or sublimation) of the immediate, impulsive, emotionally driven instinct to confront, to rupture the relationship, to "do something," with all its potential for irremediable and irrevocable loss. The ability to "frame," to see the big picture, is a factor here, as is humility. We don't acknowledge it very often, but there's an undeniable element of vanity in our need for instant emotional gratification when we are angry or feel wronged.

If, as neuroscientists like Elkhonon Goldberg have argued, wisdom can be defined in a narrow sense as an advanced form of pattern recognition by the brain, research on the role of emotion in problem solving by older adults poses some important challenges to biologists. The clear implication of the work in social cognition is that emotion is woven into memory, and therefore it, too, is part of the pattern. Wisdom, in this view, is not merely the perception of a familiar pattern of a landscape or background, but the extraction of social, emotionally salient details associated with that pattern. Thus, emotion may shape the search for, and appraisal of, the kind of information that informs judgment and decisions during a tense social confrontation. In short, emotion even shapes the kind of information we gather on our way to making what we hope will be a wise decision. In this process, we should pay attention to our inner grasshopper.

Finally, and at the risk of sounding like a broken record, the research of psychologists like Blanchard-Fields poses considerable challenge to the findings of current work in neuroeconomics and the biology of decision making. An experimental bias—a *huge* bias, from the point of view of life-span developmental psychology—is embedded in the methodology of many neuroeconomics studies because they rely almost exclusively on college-age volunteer subjects. If the life-span psychologists are right, if older adults benefit from more sophisticated emotional strategies in their behavior, then older adults may well process decision-making information (and regulate the emotions that affect such decisions) in a significantly different way than young people do. This doesn't necessarily mean that older is wiser. But it might mean that being older improves your chances of being wiser.

In 1988, in the eighty-seventh year of his life, the man who put wisdom on psychology's plate offered some further thoughts on its role in old age. Erik Erikson and his wife, Joan, who also wrote about wisdom, spoke with the writer Daniel Goleman about the increasing importance of wisdom in life. From their octagenarian perch, the Eriksons viewed the classic stages of human development, as laid out in Erikson's earlier work, as a priming mechanism for a culminating kind of wisdom in old age. Erikson located the origins of many of the traits we have been talking about—empathy, resilience, humor, humility, intuitive knowledge, altruism, generosity, an appreciation of limits—in earlier stages of life, from infancy through early childhood and adolescence and on through middle age.

"What is real wisdom?" Joan Erikson asked at one point in the conversation. "It comes from life experience, well-digested. It's not what comes from reading great books. When it comes to understanding life, experiential learning is the only worthwhile kind; everything else is hearsay."

Perhaps the ultimate expression of wisdom, however, lies in what Erik Erikson called "generativity"—taking the initiative and responsibility to share with future generations what you have learned in your own lifetime. "The only thing that can save us as a species," Erikson said in the interview, "is seeing how we're not thinking about future generations in the way we live. What's lacking is generativity, a generativity that will promote positive values in the lives of the next generation. Unfortunately, we set the example of greed, wanting a bigger and better everything, with no thought of what will make it a better world for our great-grandchildren."

Erikson's remarks on generativity nearly a generation ago raise an important question. Is there still a role for wisdom in our technologically know-it-all societies? And if so, where?

CHAPTER FOURTEEN

CLASSROOM, BOARDROOM, BEDROOM, BACK ROOM

Everyday Wisdom in Our Everyday World

*If what most men admire, they would despise,
'Twould look as if mankind were growing wise.*

—Poor Richard

DEEP INTO HIS SINGULAR ADVENTURE in parenting (he fathered seventeen children in all), Josiah Franklin faced imminent rebellion from his youngest son, who, as he entered his teenage years, refused his father's pleas to join the family trade, a small business devoted to the manufacture of soap and candles. In a similar confrontation with an older son, Josiah had asserted his authority and won the battle of wills, but he lost his son in the bargain. To the family's everlasting remorse, the youth left home, became a sailor, against his father's wishes, and eventually perished at sea.

Now came a return engagement of this battle immemorial: parent versus child, adult versus adolescent, authority versus leniency, family allegiance versus independence, tradition versus change. If domestic wisdom involves learning from emotional history rather than repeating it, then Josiah had learned his lesson well by the time he locked horns with his fifteenth child. Rather than relying on parental authority, Josiah took a shrewder, more indirect tack this time. Acknowledging his son's intense dislike of candle making, and fearing that he, too, might forever flee the family fold, the father began to accompany his son on informal walks where they could observe other tradesmen at work—bricklayers, carpenters, joiners, and the like—in the hopes of identifying a craft more to the young man's liking. Ultimately, this son ran away from home, too, but not so far; he settled on typography as a trade, and he began a thriving printing business in Philadelphia. Few people would argue, even in the age before electricity, that keeping a son is more important than the world losing another candle maker, especially when the son would later become one of the founding figures in the science of electricity. And Benjamin Franklin later appreciated his father's wisdom in this regard. "His great excellence," he wrote of his father, "lay in a sound understanding and solid judgment in prudential matters, *both in private and public affairs*" (italics added).

As Franklin *fils* implied, wisdom has an outside voice and an inside voice, assumes a role on the grandest of stages, yet can also affect the trajectory of a life still waiting in the wings. As Franklin pointed out in his *Autobiography*, his father never engaged in public affairs, and yet a steady stream of Colonial movers and shakers came to his Boston home, dined at his table, and solicited the elder's judgment in matters related to church, state, and life in general ("he was also much consulted by private persons about their affairs when any difficulty occurred, and frequently chosen an arbitrator between contending parties"). Indeed, the Franklin dinner table was an incubator of wisdom, crackling with edifying conversations leading to "some ingenious or useful topic for discourse" for the benefit of the children. The quality of the conversation was so exceptional that, in a phrase that would prick the ears of modern cognitive scientists, the Franklin children were brought up with a "perfect inattention" to the quality of the food set before them; attention, after all, is the brain's way of setting priorities and deciding what is most important. Josiah Franklin was, in short, a wise man, although no one other than his son seems to have taken public note of the fact.

The life of Benjamin Franklin is a primer of streetwise wisdom; his writings brim with shrewd observation. Franklin often distinguished between cunning (a shortcut solution to a short-term problem) and wisdom ("Be neither silly, nor cunning, but wise"), but he walked both sides of the street that separates pragmatism from virtue. His life exemplifies the many arenas of daily endeavor in which all of us, in ways "unknown to fame," can nonetheless strive for wisdom and make a daily difference in our own lives and in the lives of the people who matter most to us. For Franklin, it began with a love of learning and dedication to self-improvement, from which he derived as much nourishment as from the food at that dinner table. But he took his own prescription for wisdom in his life as an educator, entrepreneur, politician-diplomat, and even domestic partner.

Taking our cue from Franklin, these four hubs of human activity—education, business, politics, and family matters—mark places where, even today (*especially* today), we could all use a little more wisdom.

WISDOM IN EDUCATION

What signifies knowing the Names,
if you know not the Natures of Things.

—Poor Richard

Several years ago, the office of undergraduate admissions at Tufts University, in Medford, Massachusetts, began to invite prospective students to submit an optional essay that, unbeknownst to the applicants, offered a window into their potential for creativity, practicality, critical thinking, and wisdom. The questions tend to be whimsical, but with lots of rhetorical room to roam. One year, the question was "What is more interesting: gorillas or guerillas?" Another year, a question began by quoting astronomer Edwin Hubble, "Equipped with his five senses, man explores the universe around him and calls the adventure Science," and then asked, "Using your knowledge of scientific principles, identify 'an adventure' in science you would like to pursue and tell us how you investigate it."

Putting the words *undergraduate* and *wisdom* in the same sentence probably sounds like syntactic malpractice, but the Tufts program is anything but accidental. Each year since 2006, the admissions office has cooked up a new set of questions to "probe" for wisdom, and each spring Lee Coffin, the dean of undergraduate admissions, and his colleagues pore over nearly eight thousand essays. They are constantly amazed by the ability of adolescents to rise to the challenge. "I find them to be remarkably effective pieces of writing," Coffin told me.

The student essays reveal many of the qualities we have been talking about: the ability to put oneself in somebody else's shoes, the ability to see social needs larger than oneself, the ability to see the big picture, the ability to understand that situations, and truths, change with the passage of time. These abilities, needless to say, are utterly opaque in standardized testing. As Coffin put it, "In a highly subjective selection process, more information is better."

For centuries, the finest minds have lamented the moat between formal education and conspicuous wisdom. Montaigne made the distinction while ridiculing a student of Dionysius: "All the advantage you recognize is that his Latin and Greek have made him more conceited and arrogant than when he left home. He should have brought back his soul full; he brings it back only swollen; he has only inflated it instead of enlarging it."

While it is probably true (or at least *should* be true) that every institution of higher learning aims to enhance the wisdom of its students, few have made such a systematic effort to incorporate aspects of wisdom into the curriculum, the admissions process, and even into the day-to-day mission of the school than has Tufts. That is largely because Robert J. Sternberg, current dean of the School of Arts and Sciences at the university and a former president of the American Psychological Association, has put a three-decade academic career on the line and decided to test the idea that wisdom can be cultivated in young minds—indeed, *needs* to be cultivated—if the world has any hope of changing.

In a sense, Sternberg became a crusader for wisdom through the failure of the work that won him his greatest academic fame. As a professor of psychology for more than thirty years at Yale University, he developed a comprehensive theory of human intelligence. The definition, in a nutshell, argued that successfully intelligent people possessed a suite of three crucial skills—the creative ability to come up with fresh ideas, the analytic ability to decide which of those ideas were truly good, and the practical ability to convince other people of their value, thus allowing their implementation.

Around the mid-1990s, however, Sternberg began to recognize fatal flaws in his own theory, which could be boiled down to a succinct question: How is it that people who are certifiably intelligent do such foolish, and occasionally such evil, things? Sternberg tends to think of wisdom (and intelligence) as it plays out on a large public stage, so he wrestled with the untidy realization that people like Adolf Hitler and Josef Stalin satisfied all his criteria for intelligence—they were imaginative, intelligent, socially shrewd, and had the will to turn their ideas into reality. It's just that they also happened to be monsters.

"What differentiates really great leaders from not so great ones?" he said in an interview. "I looked at people like Gandhi and Martin Luther King, Jr., and Mother Teresa and Nelson Mandela—take your own pick—and if you compare them to Stalin and Hitler and Mao, they probably didn't differ much in IQ. It seemed that what differentiated them was *wisdom*." The missing ingredient, he realized, was the ability to enlist all these elements of intelligence in the service of "a common good." As Sternberg recently put it, in an essay written with colleagues Linda Jarvin of Tufts and Alina Reznitskaya of Montclair State University, "What matters is not *only* how much knowledge you have, but how you use that knowledge."

How you use knowledge, of course, acknowledges the importance of paying attention to the greater public good—which, in a sense, is a *very* old idea, if you accept the evolutionary history of altruism. But Sternberg also realized that wisdom represented a state of mind beyond standard metrics of intelligence, and this revelation forced him to see inherent failures in the educational system, in the philosophy of educational testing, and the degree to which narrow measures like IQ tests fail miserably at predicting lifetime satisfaction. (He has gone on to become one of the gurus in the wisdom field, editing the first formal academic text on the subject in 1990.)

What exactly does Sternberg mean by wisdom? He and his colleagues define wisdom—and forgive the academic-speak for a moment—as "the application of intelligence, creativity, and knowledge as mediated by values toward the achievement of a common good through a *balance* among (a) intrapersonal, (b) interpersonal, and (c) extrapersonal interests, over the (a) short- and (b) long-terms, in order to achieve balance among (a) adaptation to existing environments, (b) shaping of existing environments, and (c) selection of new environments."

Granted, this definition reads like a contract for something you'd never want to buy, but it is wonderfully supple when you parse its elements, and it should be clear that what Sternberg is talking about here applies well beyond the classroom. When he speaks of balancing short- and long-term goals, he addresses that timeless tussle between our future selves and our current desires, and the maturation that leads to sound planning and prudence. When he speaks of adapting to "existing environments," he invokes the emotional flexibility needed to cope, whether with a difficult passage in a marriage or daunting financial circumstances. When he speaks of "shaping" an environment, he suggests that wisdom can act as a change agent, using decisions and action to remake a domestic or business situation. And when he speaks of "selecting" a new environment, he also admits that there are situations where wisdom's agent seeks a change of scenery, whether it is leaving an abusive relationship or deciding to change jobs. Most of all, wisdom is a balancing act, a kind of spiritual gyroscope that seeks and requires equilibrium in the face of constantly changing forces and interests.

Sternberg's conception of wisdom infiltrates virtually every aspect of modern life and represents, as he puts it, "a complex model accounting for real behaviors in real contexts." It also dresses up a lot of modern neuroscience in humanistic garb; the behaviors he describes pass, by other names, as temporal discounting, framing, exploratory and exploitative behaviors, emotional regulation, and "meta-wisdom."

The grandness, the *messiness,* of Sternberg's definition of wisdom smothers the conventional metrics of intellectual achievement. As he puts it, the current emphasis on test results, from SATs to GREs to standardized math and English measures in primary and secondary education, is "orthogonal to wisdom." And as he ruefully admitted, Tufts University used to be part of the problem, its business school having bestowed on the world of finance both Kenneth Lay and Andrew Fastow—two very bright graduates who, as executives at the Enron Corporation, were also the architects of one of the greatest corporate scandals and collapses in the history of business. "We have constructed an educational system," Sternberg says, "to produce people with skills to lead us in exactly the direction we don't want to go."

What would an education that valued wisdom look like? Sternberg and his colleagues have been trying to figure that out. It hasn't gotten much attention, but it's one of the most interesting experiments in higher education.

In the Tufts system, teachers are asked to do something that hearkens back to the original Academy of Plato and Lyceum of Aristotle: serve as role models of wisdom. In a Socratic, show-rather-than-tell approach, teachers try to elicit new habits of thoughts in their students: how to balance competing interests in everyday decision-making tasks, how to incorporate one's moral and ethical values into one's thought processes, how to think "dialogically" (taking an other-centered approach that attempts to understand multiple viewpoints), how to think "dialectically" (understanding that a solution that is right at one time and place may well be wrong when circumstances change), and how to become self-conscious in a positive and enlightening way, monitoring one's own thought processes and decisions through a lens of wisdom. In a wisdom-based approach to teaching, Sternberg has written, teachers "will take a much more Socratic approach to teaching than teachers customarily do" and "students will need to take a more active role in constructing their learning."

And you're never too young to start incorporating wisdom into learning. While still at Yale, Sternberg and some colleagues took a crack at designing a curriculum for middle school students called "Teaching for Wisdom," with the aim of nurturing the development of wise and critical thinking. In 2002, the Saddle Brook School District, in northern New Jersey, began a trial implementation of the program; the experiment lasted all of one semester, but teachers there still rave about the impact of the approach. The results were "fabulous," according to Marilyn Hamot Ryan, the coordinator of gifted programming in the Saddle Brook School District. "For me, to work with kids like this, knowing so much beyond the usual academics, getting kids to think outside the box—it was fantastic," she said. "We took some of the basic philosophy of that pilot program and incorporated it into the general curriculum."

The "Teaching for Wisdom" philosophy was applied to the study of history, science, literature, and foreign language. In one educational exercise that particularly resonated with young people (and is still used in Saddle Brook schools today, according to Ryan), students read Ben Franklin's *Poor Richard's Almanack* and then created personal notebooks full of maxims they had learned in the course of their own lives—and from interviewing grandparents and other older people. As Ryan put it, "It's a great way to motivate the kids, to get them to think beyond themselves."

One small pilot program does not an educational revolution make, but it does raise a big question: Why is there so little interest among professional educators in the cultivation of wisdom in the intellectual and emotional development of our children? Is it because it is much easier to test for algebra and grammar than for incisive, other-centered thought? Sternberg is adamant in his belief that cultivating wisdom with the public good in mind is critical to the future success of our society as we struggle to deal with issues like global warming, weapons of mass destruction, and social justice. "In the end, wisdom is the only thing that will save us," he says. "It's all about doing the right thing."

WISDOM IN THE WORKPLACE

Industry need not wish.

—Poor Richard

Wisdom in the workplace typically implies two separate, and rather distinct, areas of wise behavior. One is the wisdom of corporate decision making and all the things that go into skilled business judgment: knowing what information to use in decision making, creating a culture of knowledge in order to acquire that information in a timely fashion, assessing it in both short- and long-term frameworks, and, of course, reaping the financial rewards that come with such shrewd financial choices. In many cases, this involves not so much neural circuitry as just plain hard work. As Poor Richard said, "Diligence is the Mother of Good-luck."

Several recent trends in business theory, however, seem to be modeled, intentionally or accidentally, on the brain's own decision-making process, where fact-based knowledge (or experience) competes with emotional urges (intuition or hunches). Tom Davenport, professor of information technology at Babson College in Massachusetts, specializes in "business intelligence," which he describes as the "systematic use of information about your business to understand, report on and predict different aspects of performance." In interviews, Davenport argues that leadership is the most important factor in cultivating this organizational thought process, citing as examples Jeff Bezos of Amazon.com, Inc., Gary Loveman at Harrah's Entertainment, Inc., and Reed Hastings of Netflix, Inc. (all of whom, incidentally, had more than casual training in mathematics). "It's how committed are a company's senior executives to fact-based and analytical decision-making and the whole idea of experimentation as a way to learn rather than doing it out of gut feel or intuition."

But shrewd business acumen may be less a fight between our inner grasshopper and ant and more a matter of relying on one of the fundamental neural pillars of wise behavior: emotional regulation. In a famous profile of investor Warren Buffett, financial writer Joseph Nocera kept returning to two qualities that, to his mind, accounted for "Saint Warren of Omaha's" phenomenal success. One was Buffett's discernment. If accounting is the language of business, as Buffett has said, he is unusually fluent; as Nocera put it, "he finds meaning in numbers that the rest of us don't." Like a lot of forms of wisdom, Buffett's financial wisdom is built upon a foundation of expert knowledge. But the other quality has nothing to do with intelligence or knowledge. Nocera writes that "truly great investing requires a temperament that very few people have," and he goes on to point out that Buffett "is never, ever ruffled—which I think is a key to great investment: His judgment seems never to be affected by emotion."

So the current edition of the business sage is coldly analytical, empirical, willing to learn from mistakes, and able to control—or, as a Buddhist might say, master—the emotions, right? Well, not so fast. The other, less appreciated aspect of corporate skill is social wisdom—a skill that is captured by the obligatory (and usually derisive) designation known as "human relations."

So much of our physical and psychic energy is depleted by conflicts, stress, and interpersonal tensions in the workplace, and yet we tend to measure "economic wisdom" solely by the standard marketplace yardsticks—sales, profits, dividends, and so on. The bottom line is always financial, never reckoned in terms of job satisfaction, sense of personal fulfillment, or fully tapping a group's entrepreneurial or innovative potential. But as the toll of this dissatisfaction becomes everywhere apparent, perhaps there is another way to exercise wisdom in the workplace—by identifying and implementing ways to maximize the potential of both an organization and its employees. How many companies have foundered when their leadership, while fundamentally successful according to the ledger book, has taken disastrously unwise actions?

As almost any manager will tell you, the major part of a manager's time is spent refereeing, manipulating, soothing, inspiring, cajoling, and otherwise massaging social relationships in the workplace, creating the kind of corporate environment that maximizes the performance of the team while minimizing the social and emotional frictions between individuals that often drag upon performance. As in a marriage, it's not simply what you say in the office, but how you say it, and when, that determines whether the message gets through. And, in the kind of complex and contingent behavior that social wisdom so often demands, wisdom in the workplace sometimes requires the extra effort of treating people differently, as individuals, in order to keep the entire group unified around the greater common goal.

If anyone thinks this customized approach is excessively time-consuming or economically inefficient, there have been few people better at maximizing group performance under exceptional pressure than that pinstriped guru Casey Stengel, the Hall of Fame baseball manager who won seven world championships with the New York Yankees over twelve seasons. "The secret of managing," Stengel explained in an oft-quoted remark, "is to keep the guys who hate you away from the guys who are undecided." In this case (though not always), what's true of the locker room is also true of the boardroom and the workplace.

Confucius is the unsung sage of the locker room, the cubicle, anyplace where workers convene. Countless observations in the *Analects* can be read as a guide to wise, discreet, successful behavior in the workplace; indeed, in his attempts to create a code of proper conduct in a feudal society, Confucius produced a marvelous guide to bureaucratic survival, especially for those in the tiers of middle management. Part of his advice had to do with personal behavior, part with interpersonal relations, but all of it contributed to a sense of the meritocratic, occasionally self-denying, often humble, and always shrewd way of negotiating relations with both bosses and underlings in any hierarchical structure.

As he told the disciple Jan Yu at one point, "Get as much as possible done by your subordinates. Pardon small offenses. Promote men of superior capacity." When asked how he discerned such superiority, Confucius replied, "Promote those you know, and those whom you do not know other people will certainly not neglect." He also held executives ("gentlemen," in his formulation) to a higher standard: "A gentleman takes as much trouble to discover what is right as lesser men take to discover what will pay."

Is compassion compatible with good business? Recent studies suggest (at least prior to the financial crisis) that a bighearted, right-minded, other-centered business may, paradoxical though it might seem, be an excellent sign of a strong, healthy business. Researchers at New York University and the University of Texas recently published a study showing that companies with a higher commitment to charitable giving and philanthropy—among the corporations cited were Johnson & Johnson, Pfizer, PepsiCo, Nike, Kimberly-Clark, and Avon—consistently ran good businesses. The NYU-UT study concluded that "doing good is apparently good for you."

Indeed, precisely that metric—doing good for the community—was the key marker for a mutual fund run by Dover Management of Greenwich, Connecticut, which invested in companies known for charitable giving, with the thought that only financially healthy companies can afford to be generous. "A philosophy from top management filters throughout the organization," Michael Castine, then president of Dover, told *The Wall Street Journal* in the fall of 2007. In one recent year, this do-gooder fund even beat returns on the S&P 500 index.

An other-centered corporate personality is usually shaped by the CEO, Castine told me when I spoke with him, and it has long been a feature of large, well-capitalized companies. "John D. Rockefeller spent every waking hour making money," Castine said, "and he put as much effort into giving it away as making it." Companies like Nike and Avon made charity part of their business identity, but they also shrewdly tied their philanthropic initiatives to their brand—Nike gave away shoes; Avon focused on breast cancer awareness. There may even be a cultural component underlying corporate largesse. Companies based in Minnesota have historically been especially generous about corporate philanthropy, according to Castine. "Target, Dayton-Hudson, General Mills, Pillsbury—all have been huge givers over the years," he said. "There's a lot of people in Minnesota from northern Europe, from Scandinavia, and all those places have a big commitment to giving back to the community." But this kind of corporate magnanimity is also fragile. By the time I caught up to Castine in the fall of 2008, he told me that Dover had closed the fund—"before the market tanked, but when the hand-writing was on the wall."

Finally, a wisdom-related business issue that is often overlooked is the quality of the workplace. As a science writer, I have always been intrigued by the sociology of the lab group—which labs are consistently productive and innovative, how they function as social organisms, what role the lab chief plays in establishing that group ethic, and how various groups handle setbacks and failure. Two decades ago, while reporting for my first book, *Invisible Frontiers,* I became fascinated with the "personality" of the three laboratories that essentially pioneered the biotech industry. These three labs competed in a high-stakes race to be the first to clone a human gene—in this case, the gene for human insulin, which in 1983 became the first genetically engineered drug to reach the market.

One lab, headed by Walter Gilbert at Harvard University, had the public reputation of being fiercely competitive, a place where you were either in first place or no place in a scientific race. Yet interviews with both postdocs and graduate students revealed a tight-knit, hardworking, and free-spirited group of researchers who were given the freedom to succeed (and fail), as well as unusual latitude to chide and criticize their lab chief (a satiric photocopied lab newspaper called *The Midnight Hustler* became a vehicle for poking frequent fun at Gilbert and anyone else in the lab who suffered lapses in humility or tried to claim inappropriate credit). Despite a spectacular experimental failure in England that doomed the Harvard team's chances in the insulin race, Gilbert remained emotionally steady and philosophical; he went on to win a Nobel Prize and cofounded the biotech company Biogen.

The second group of researchers, at a fledgling company named Genentech, also enjoyed unusual freedom, unusual support (of both the financial and emotional sort), and unusual camaraderie. Working around the clock with a minimum of hierarchical control, they went on to win the race. It is perhaps no accident that now, some thirty years after its initial triumph in the insulin race, Genentech continues to rate near the top in corporate surveys of good places to work.

The third lab in the race, based at the University of California at San Francisco, presented an entirely different story. It suffered from legendary internal competitions and jealousies, secretiveness, and acrimony, which still haunt its former members; one of its lab chiefs was hauled before Congress in 1977 to testify about violations of federal guidelines for recombinant DNA experiments, and some of the conflicts have produced litigation that has continued more than three decades after the historic experiments. Just as business has its famous case studies, so does science, and these examples remain vivid cautionary tales in the scientific community. I recently heard a well-known Nobel laureate referring to the three insulin-race labs in the context of a conversation with colleagues about "happy labs" and "unhappy labs." The tone for both, scientists acknowledge, is always set at the top.

Almost everyone with whom I spoke from the Harvard lab also remembered a maxim that Wally Gilbert constantly preached to young scientists, an idea that is just as applicable to excellence in business as to excellence in research. It takes as much time to solve a hard problem as an easy one, he would say, so you might as well take on the challenge of a hard, important problem. Industry, indeed, need not wish.

WISDOM AT HOME

When Man and Woman die, as Poets sung,
His heart's the last part moves, her last, the tongue.

—Poor Richard

For all his practical wisdom, Ben Franklin never quite got up to speed on domestic relations, at least as far as his alter ego, Poor Richard, goes. The *Almanack* is full of cringing, sexist doggerel like the lines above, which merely reiterate a point often made by psychologists: It's hard, even for an exemplary and virtue-minded mortal like Franklin, to be wise all the time. Although one could easily fill a book exploring ways of exercising domestic wisdom, I don't plan on trying, except to note that our models for this kind of exemplary behavior are almost always close to home and, in Franklin's apt phrase, "unknown to fame." With that in mind, allow me to make a few observations, beginning with lawn care.

When I was growing up, my father spent what seemed like an inordinate amount of his free time weeding the lawn. I can still picture him in his shorts, bent over in a way that exposed the ghastly leg scar from a bullet wound he suffered in World War II, the *snip-snip* sound of the clippers like some macabre suburban metronome, marking the death of another blade of grass and the passage of another misspent afternoon. When I was a teenager, this always struck me as a waste of precious domestic time.

Now, in my midlife career as a weekend weeder, I have come to realize that he probably found this routinized, mindlessly repetitive activity a kind of backyard mindfulness—very calming, a way to flush his mind of daily concerns, a ritual that allowed him to focus on, reframe, and think through more important things. I realized this one morning many years later when I'd had a sharp exchange of words with my wife, a disagreement leading to tears and hard feelings. I retreated to our backyard garden to pull some weeds (always a remunerative use of one's time when there are deadlines to meet and bills to pay). As I concentrated more and more on the weeds, my anger subsided, to the point that I realized who had precipitated the fight (me) and what had prompted it in the first place (as usual, a displaced and unspoken resentment). I stood dumbfounded for a moment, holding a bouquet of weeds in my hand, and then went back into the house and apologized. My wife was more delighted than if I'd brought her flowers.

We aspire to domestic wisdom in our relationships with parents, children, spouses, and loved ones. It is a form of wisdom that does not require a large public stage, yet its impact can affect the trajectory of our children's lives or the fate of a marriage. Vivian Clayton first became fascinated by the idea of wisdom while observing the actions of her father as a family patriarch. She recalls that he had the ability to discern decisions that needed to be taken right away and those that required more deliberation; even more, he always had the ability, she says, to see clearly what would be in the best interests of the entire family. These are timeless values about the "pragmatics" of life that are as true (and necessary) today as they were when Clayton was growing up in the Brooklyn of Jackie Robinson and egg creams.

For most of us, domesticity is a series of unannounced pop quizzes on interpersonal wisdom and social grace—whether we are dealing with partners or soul mates, immediate or extended family; whether we live in a dorm suite or have three generations of kin under one roof. Regardless of the temperature settings on these various relationships, they all require cooperation, flexibility, and, in the jargon of our primate friends, social

grooming. So much of wisdom hinges on decisions that involve our relationships with others that almost everything we've been talking about in
terms of wisdom, neurologically and psychologically, eventually curls
back to a discussion of social intelligence.

My little retreat to the garden forced me to call on three qualities crucial to domestic social grooming: emotional regulation, taking the point
of view of another, and humility, which is the emotional lubricant necessary to squeeze out those difficult apologies. Keeping an even temper in
moments of challenge—or at least retreating to a reliably calm place
where it is easier to reset your emotional dial—is obviously key to maintaining emotions on an even keel. As Laura Carstensen's research at Stanford has consistently shown, older adults seem particularly good at
quickly letting go of negative emotion, and their motivation for doing so
seems very clear. They value social relationships more than the ego satisfaction that would come from rupturing them. If judgment ultimately
depends on the quality of our evaluation, these are wise values indeed.

Second, a domestic relationship often seems like a game that is paradoxically won by losers—and I mean that in the most complimentary
way. People who insist they are always right, who insist on winning every
argument, who insist they are never to blame (and I speak with some
authority on this point), often feel ratified by the assertion of their power
in intimate relationships. But in a world of ongoing social interaction, as
in the serial repetitions of games in game theory, people who insist on
winning at all costs, whose self-interest trumps sociality, and whose
greed (financial or emotional) exceeds the bounds of fairness end up
playing solitaire. A famous Sicilian proverb holds that "the man who
plays alone never loses," but in the social reality most of us experience, it
also means he never wins. Apologies are social and emotional loss leaders; they require capitulation, and capitulation is impossible without
humility.

One other quality is essential to a long-term relationship: patience.
When I make a mistake as a parent—and I make them every day—it
almost always starts with impatience. Tone of voice, pained expression,
the body language of exasperation—you can read me like a book, and
my children have become speed-readers. This is another facet of emotional regulation, of resisting the instant gratification to voice that surg-
ing (and often justified) dissatisfaction. Now that I'm on the other side

of the parental divide, I sometimes ruefully admit that being a child is often akin to working for a meddling, demanding, and never-satisfied boss, and it's very easy for the *habit* of the relationship to settle into a curt emotional shorthand. The mental discipline and strength required for patience is more wearying than an hour at the gym, yet nothing is more essential to the day-to-day health of a family.

How do we learn to be patient? Part exertion and part example, I suspect. My father was an extremely patient man, although I gave him and my mother plenty of reason for disappointment and exasperation. I remember in particular a conversation late one evening when I popped into my parents' bedroom to announce that I planned to drop out of college—the next day. My mother reacted with wounded outrage and invective (which, of course, I've never forgotten and barely forgiven—things said in anger leave their own emotional paper trail).

My father, after allowing the verbal dust from this unexpected bombshell to settle, pointed out that I'd just been accepted into a special program at school and it would be unfair to all the other applicants not to give it a try at least. I knew about as much about wisdom at the time as any eighteen-year-old, which is to say nothing, but I realize now just how brilliant his response was, as beautifully digressive as one of Josiah Franklin's neighborhood walks. It betrayed no impatience; it steered the conversation away from family dynamics; it appealed to a sense of fairness and responsibility, not to my family (a tactic utterly wasted on most eighteen-year-olds), but to my peers.

We all went to bed that night, and the next day I went back to school. These are life-defining moments, and if we're lucky (and I was), there will be at least one wise person in the room.

WISDOM IN POLITICS

Kings have long Arms, but Misfortune longer:
Let none think themselves out of her reach.

—Poor Richard

In 2007, the columnist David Brooks reviewed a book entitled *The Political Brain,* by Emory University psychologist Drew Westen, in which he sharply criticized Westen's thesis that emotions drive voter choice. Brooks concluded, "The best way to win votes—and this will be a shocker—is to offer people an accurate view of the world and a set of policies that seem likely to produce good results." In 2008, Brooks wrote a column called "How Voters Think." This time around, he argued that voters in fact do not make "cold, rational decisions about who to vote for." Rather, voters make "emotional, intuitive decisions about who we prefer, and then come up with post-hoc rationalizations to explain the choices that were already made beneath conscious awareness."

No mainstream commentator has been more attuned to the monumental social implications of modern neuroscience than Brooks, so he is a kind of genial avatar for our struggle to understand the ramifications of this new science and what it means for the possibility of wisdom in our politics. From my vantage point, the second Brooks is much closer to the mark than the first—unconscious emotion drives much of our decision making, political or otherwise. But as we explore the possibility of wisdom in politics, let's begin by acknowledging that we are talking about two quite distinct domains of political wisdom, with two distinct burdens of responsibility: wise leaders and wise voters.

Once upon a time, it was not an automatic oxymoron to put the words *leader* and *wise* in the same sentence. Solon, one of the Seven Sages of the ancient world, burnished his deserved reputation for political wisdom when he instituted massive reforms in the entire structure of Athenian law—and then shrewdly took a ten-year "leave of absence" (Plutarch's phrase), so that the "special interests" in ancient Greece could not lobby him for changes. Even in that golden age, the folly of both legislators and voters was painfully apparent. Solon's friend Anacharsis, Plutarch says, "was amazed to find that in Greece wise men spoke on public affairs, but fools decided them."

Since liberal democracies are very much the grandchildren of philosophy, the histories of wisdom and politics are deeply intertwined. Indeed, in the view of some, the attainment of wisdom moved from an individual calling prior to the eighteenth century to a grand endeavor of nation-states, fueled by the ideals of both the French and the American revolutions. "The attainment of justice and happiness were to become the art of organizing a just society that delivers happiness to its members through collective justice," Jean-François Revel said. "At that moment, the whole ethical branch of philosophy was reborn in the form of political systems." The price of this political transition, especially in the twentieth century, has been not only serial disillusionments with fascism, Marxism, socialism, and, perhaps, capitalism but spiritual disillusionment on a personal level. As Revel argued (as only the French tendentiously can), "So there are no more individual ethics, and there's no longer any personal quest for wisdom."

There's not much political wisdom around, either. Although commentary on wise (or unwise) political leadership is almost as old as literacy itself, few have addressed the issue more bluntly and shrewdly than the historian Barbara Tuchman. In a 1979 speech to the United States Military Academy, later republished as a much-cited essay, Tuchman gave a brisk seminar on the history of what she called "An Inquiry into the Persistence of Unwisdom in Government." It's an authoritative analysis of human limitation, and it should be required reading for every would-be politician (and, indeed, every about-to-be voter).

To be sure, Tuchman offered some role models of wise leadership: Pericles, Julius Caesar, Marcus Aurelius, Charlemagne, and George Washington all find a place in her pantheon. But her essay primarily addresses the issue of governance through the lens of human folly; though it was written long before the modern era of neuroscience, her insights resonate with many of the traits cognitive scientists have been studying.

Unwisdom in leadership, in Tuchman's view, often boils down to a failure of character, a personal estrangement from the virtues we associate with wisdom: a sense of fairness, humility, emotional regulation, deliberation. Hence, "wise policy can only be made on the basis of *informed,* not automatic, judgments." Lack of self-control is also a feature of unwisdom, with an added gender component: "Government remains the paramount field of unwisdom because it is there that men seek power over others—only to lose it over themselves." Self-delusion? Government "excites that lust for power that is so subject to emotional drives, to narcissism, fantasies of omniscience, and other sources of folly." Tuchman concludes on a forlorn note, suggesting that it's probably worthless to address political unwisdom by trying to educate the governing class; better to concentrate on educating the public. She believed that "fitness of character is what government chiefly requires. How that can be discovered, encouraged, and brought into office, I have no idea."

Back in 1979, Tuchman lamented the way money and "image-making" had begun to manipulate the elective process. One can only imagine the horror with which she might scan the current landscape, where political campaigns have recruited neuroscientists to help them understand, and shape, voter preferences. As *The Wall Street Journal* reported during the 2007–2008 presidential primary season, several candidates enlisted the aid of clinical psychologists and other campaign consultants who use methods of neural measurement, including fMRI scans, to eavesdrop on the brains of potential voters as they responded to campaign speeches and platforms. These new neuroscience consultants essentially told their clients (John Edwards, Mitt Romney, and George W. Bush reportedly have used them) how to short-circuit reason and appeal to subconscious biases. Some of these approaches aspire to illuminate the "neural networks" that are engaged when voters decide who they like. The neuroscientific political consulting group EmSense Corporation, in California, for example, monitored brain activity in potential voters while they watched televised political debates; press accounts described the detection of "a pronounced shift in activity in their prefrontal lobes" when one candidate proclaimed his ability to get things done. When another candidate suggested that a withdrawal of American troops from Iraq would invite a sanctuary for terrorists, researchers noted a concurrent rise in adrenaline (indicating a rise in the fear response) in prospective voters.

If you believe that unconscious, emotion-driven intuition leads to good political judgment, then these tactics are just a more sophisticated form of voter education. But if you believe George Lakoff, the UC Berkeley neurolinguist (and political liberal), it is a deeply mischievous form of neural manipulation. Lakoff has written about the biology of political decision making in his book *The Political Mind,* where he argues—backed by decades of neuroscientific and psychological experimentation, especially the Nobel Prize–winning work of Daniel Kahnemann at Princeton University—that political decisions can be cognitively "framed" by the way the choice is presented to voters. And often, this framing is accomplished by the use of loaded, focus-group-tested words—some of them subtle, like *tax relief,* and some of them explicitly visceral, like *Willie Horton* or *Swift Boating.* In Lakoff's view, the language of politics, whether in speeches or advertisements, is a meticulously composed mosaic of code words, inflections, and implicit associations. As in Jonathan Haidt's experiments with hypnosis, these code words are designed to detonate unconscious emotional reactions, especially fear and disgust.

As he recounts in his book, Lakoff sometimes challenges his students in a class on semantics with a simple task: Don't think of an elephant! The point is that, once the word is invoked, it automatically triggers a neural frame; you can't *not* think of an elephant. Lakoff writes, "It's not just the word 'elephant'; it's all words. And it's not just one frame that's activated unconsciously and automatically by words—it's a whole system of frames and metaphors. The more that system is activated, the stronger its synapses become, and the more entrenched it is in your brain—all without your conscious awareness. That is why the conservative message machine, operating over thirty-five years, has been so effective." It is also why, as Matt Bai chronicled in a fascinating *New York Times Magazine* article, the Democratic Party rushed to adopt Lakoff's ideas on framing in its recent campaigns.

If there is a single message that has emerged from recent cognitive neuroscience, it is that the lightning-quick, emotional response system, while it undoubtedly kept humans alive in the fullness of evolutionary time and plunges us into a bath of invigorating neurotransmitters, often preempts a more deliberative, rational decision-making process. Yet that slower, newer, more rational part of the brain—the ant clinging for dear life to the back of the grasshopper, in George Ainslie's memorable phrasing—is essential to foresight, planning, and the kind of delayed emotional reward we associate with enlightened (and, yes, often unpleasant) political leadership. If we believe deliberation is useful to wisdom, then just about every tactic in the modern political campaign playbook seems designed to short-circuit (*neurologically!*) political thoughtfulness.

And neuroscientists are claiming to understand the brains of voters in ever greater detail. Brief (and controversial) reports in the scientific literature purport to reveal the neurocognitive traits of liberalism and conservatism. Marco Iacoboni of UCLA, one of the leading figures in "mirror neuron" research, got the ball rolling in 2004 with a small experiment that purported to show empathic patterns of brain activity in registered Republicans and Democrats who underwent MRI scans while looking at photographs of the three main presidential candidates (Bush, Kerry, and Nader). Many people know about these findings because they were prominently reported in *The New York Times;* many fewer know that the experiments were ultimately inconclusive, as even Iacoboni admits in his book, and failed to show what was widely reported.

No domain of neural politics seems beyond the reach of neuroscientists. In the fall of 2007, David M. Amodio of New York University and his colleagues gave us a neurological picture of a political brain steeped in ideology. In brain-scanning experiments, self-described conservatives proved to be more persistent and wedded to habit when confronted with conflict. By comparison, self-described liberals showed greater activity in their dorsal anterior cingulate, a part of the brain that monitors conflicting environmental information, suggesting they were more sensitive to cues in the environment that required a response other than habit. If you recall some of the standard components of wisdom as defined in the recent psychological research, the study's conclusions—that liberals "are more responsive to informational complexity, ambiguity and novelty"— echo some of the fundamental premises of wisdom described by the Berlin Wisdom Project.

Does this mean liberals are wiser than conservatives? Of course not. It was a small, initial study with the interesting but daunting ambition to "elucidate how abstract, seemingly ineffable constructs, such as ideology, are reflected in the human brain." But it also marks a new, potentially problematic invasion of neuroscience into contemporary political decision making.

Finally, emotional intuition has its blind spots and is not nearly as flawless as science writers, and sometimes scientists, would have you believe. In the 1980s, American voters developed a genuine emotional attachment to Ronald Reagan, but that emotional attachment may have blurred a more detached and discerning assessment—not of his politics, but of his health. In a startling passage of his 2005 book, *The Wisdom Paradox,* neuroscientist Elkhonon Goldberg claims to have spotted unmistakable signs of early Alzheimer's disease in Reagan while he was still in office, simply by watching several public appearances. Since Goldberg assesses dementia for a living at New York University School of

Medicine, this can't be dismissed as an armchair diagnosis. Watching Reagan nod off during George H. W. Bush's inauguration, Goldberg said to himself, "Brain stem gone." He added, "I was convinced that a significant portion of Reagan's second term had taken place in the shadow of his slippage toward early dementia." How good a political guide are our emotions if they prevent us from discerning a politician's deepening cognitive deficits (or any other serious flaw, for that matter)?

Is it any wonder that there are no more statesmen, no more village elders in the world of politics? Not because voters don't want it; people *yearn* for political wisdom and guidance. It's just that a system devoted to honing emotionally charged catchphrases and code words essentially creates a political culture built on a foundation of fear. The pity is that neuroscience—to the extent it can say anything credible about human decision making in a political context—is helping politicians speak to the impulsive, impatient, reactive grasshopper in our brain rather than to the reflective, prudent, future-looking ant. Is it any wonder that political leadership in the twenty-first century seems so unwise?

To those who embrace the idea that emotion reliably helps us attach value to political judgment, it's worth considering one other point of Tuchman's essay. The ultimate failure of the French, Russian, and Chinese revolutions, she wrote, can be laid to "too much class hatred and bloodshed"—too much political emotionalism, in short, which precluded "fair results or permanent constitutions." The persisting genius of the American system, by contrast, resides in its structure of checks and balances (acknowledging the imperfections of human nature) and its guarantee of civil rights (a commitment to social justice). "Not before or since, I believe, has so much careful and reasonable thinking," Tuchman wrote, "been invested in the creation of a new political system."

So we crave wisdom in the world of business, yet manufacture an economic collapse out of self-interest (greed) and a galling lack of emotional regulation (panic). We understand how important wisdom is to our domestic peace and happiness, yet we can't break the powerful neural rewards of habit, vanity, self-indulgence, and immediate gratification. We desperately need wise political leadership, yet the enormous economy of politics is bankrolling refinements in appeals to the oldest, most fearful, and most impatient part of our brains. Everyone claims to want more wisdom in our educational process, yet our current metric for excellence is the same "knowing the Names" that Benjamin Franklin derided more than two centuries ago.

One of the most dispiriting messages buried in the literature of human wisdom is the observation that there have been times during the course of human civilization when a barbarous, or merely frivolous, historical culture simply made it unwise even to appear to be wise. Socrates died in such a time, Confucius recognized the danger of such times, and Montaigne lamented the implications of such historical moments. The theme common to all these moments is the withdrawal of wisdom.

"When the Way prevails under Heaven," Confucius advised, "then show yourself; when it does not prevail, then hide." Montaigne counseled a similar withdrawal from public affairs (what he called "political government") when they had been fatally corrupted by incompetence or lack of imagination. In the essay "Of Custom," he wrote, "the wise man should withdraw his soul within, out of the crowd, and keep it in freedom and power to judge things freely."

Heraclitus withdrew even more dramatically. When he resigned his kingdom in disgust and was seen playing with children in front of the temple, the Ephesians ridiculed him for childish behavior. Heraclitus replied, "Isn't it better to be doing this than to be governing the affairs of state in your company?" Politics makes strange bedfellows; true wisdom sometimes prefers none.

CHAPTER FIFTEEN

DARE TO BE WISE

Does Wisdom Have a Future?

*Every moment instructs, and every object: for
wisdom is infused into every form. It has been
poured into us as blood; it convulsed us as pain; it
slid into us as pleasure; it enveloped us in dull,
melancholy days, or in days of cheerful labor; we
did not guess its essence, until after a long time.*

—Ralph Waldo Emerson, "Nature"

*Well begun is half done. Dare to be wise. Get
under way!*

—Horace, *Epistles*

NOWADAYS, WHENEVER I PAUSE long enough to step back and think
about our frenetic, postmodern, quasi-apocalyptic, multitasking, dual-
income, picking-up, dropping-off, stopping-by, hyphenated, bifurcated,
emotionally hectic, intellectually overwhelming, economically challenging, and spiritually benumbed lives, I'm always left with a simple question. Is there a place, a *real* place, for wisdom in our world?

Let us begin with a thought experiment. Let us reframe everything
that you've read up to this point and put it into a bigger, blacker box.
Imagine that there really is no such thing as wisdom, except as a kind of
behavioral and spiritual illusion that keeps drawing us forward, like
philosophical zombies, on a seemingly purposeful but, in reality, stum-
blingly random walk into the future. Imagine that wisdom means so
many different things to so many different people that it ceases to repre-

sent anything but the most superficial of core values; imagine further that those values, far from being timeless and fundamental virtues, are instead transitory cultural tics, products of their specific times and tribes, so that wisdom may be nothing more than a mirror beveled in a way to flatter its serial beholders. How convenient, for example, that in Renaissance Italy, coincident with the rise of the great banking families of Florence in the fifteenth century, the most astute scholastic minds decided that a fundamental prerequisite for wisdom was wealth, just as those in our era accepted until recently that greed, in the form of economic self-interest, was a virtue.

Finally, imagine that those core values—a maxim about patience, say—have been commodified and repackaged by a consumer economy that virtuously embraces bits of wisdom, as long as they are short and compact enough to fit on a coffee cup or a T-shirt. This may sound irremediably cynical, but not long ago, in the gift shop of the New York Public Library (as wonderful a temple to accumulated human knowledge and intellect as exists on the planet), I found a refrigerator magnet for sale that read, "Better to light a candle than to curse the darkness—Confucius." Is this a way of preserving pearls of wisdom in our contemporary, mass-produced form of amber, or have economies of scale managed to turn the hard-earned coin of aphoristic wisdom into one more kind of widget?

So consider this possibility: Perhaps we have been kidding ourselves all along, ever since Socrates identified the quest for wisdom as something worth dying for. Maybe we have wildly overestimated the value of wisdom, and are due for the philosophic equivalent of a painful deleveraging. What if it is nothing more than an old-fashioned virtue for old-fashioned times, where during the better part of human existence (certainly that significant chunk before Gutenberg and Gates) there was no such thing as collective wisdom, no amalgamation of human experience and insight that could act as the foundation for our decision making, and indeed no urgency to make decisions?

Perhaps we are merely comforted, and always have been, by the *idea* of wisdom—the idea that there is a higher level of human excellence and discernment to which we might all aspire. Perhaps our hunger for sages, whether they appear among us bug-eyed with biblical length of beard or adorned with the prophetic jangle of crystals and me-babble, is simply a cult of personality and reflects a human desire to suborn ourselves to forceful, decisive, seemingly *knowing* charismatic figures, a desire to be rid of the hard work of deliberation and choice (not to mention personal responsibility), to resign ourselves (with secret relief) to the more easeful role of followers and beholders.

And let's face it: Who in his or her right mind would want the burden of apparent wisdom, given history's long-running rap sheet against wise men? The role of sage, in any age, deserves hazardous-duty pay. They sentenced Socrates to death and crucified Jesus. Nobody would hire Confucius, and nobody could protect Martin Luther King, Jr. They mocked Pericles and despised Churchill and assassinated Gandhi. Even Oprah's ratings plunged (when she dared to promote the political fortunes of a relatively unknown young man named Barack Obama). If there were actually a wise person in our midst today, she would be dissected, eviscerated, macerated, and ridiculed on cable television and talk radio before we ever had a chance to figure out if she was truly wise or not; more likely, we wouldn't even recognize her as anything other than an object of social scorn—a vagrant, a crank, a disturbed personality. At his trial, as Karl Jaspers memorably reminded us, Socrates "ironically explains that since no one who speaks frankly and openly to a crowd is sure of his life, a champion of justice who wishes to remain alive even a little while had better speak only to individuals." So much for the wisdom of crowds.

And yet, you can feel an intense, almost palpable desire for wisdom in ordinary situations, among ordinary people, a hunger for any excuse to raise their game and reclaim a better self.

Why is wisdom so important to our everyday lives, and why is it so hard to achieve? For one thing, wisdom usually involves interpersonal negotiations, and these are always nettlesome—not just in the traditional domains of wisdom, such as the accounts of biblical justice and adjudication, but also in the less formal but more common setting of groups wrestling with problems and conflicts, whether the group is a family, a class, a congregation, a sports team, a committee, a work cohort, you name it. In all these dilemmas (by definition problematic), we yearn to be wise; we yearn to benefit from the wisdom of others; we yearn to find ourselves in situations that encourage and strengthen our tenuous hold on wisdom. Yet the path to wisdom is rarely clear and, as

Montaigne observed, each individual comes to her or his wisdom by a different route.

If there is a truism that crosses all cultures, from the *gen* of Confucianism to the loving-kindness of Jesus, it is that wisdom does not come easily. To paraphrase Shakespeare's famous maxim, some of us (a very few) are born wise, some become wise, and some have wisdom thrust upon them. Unlike greatness, however, the demand for wisdom is thrust upon us on a daily basis, in matters momentous and mundane, in settings as private as the bedroom or as public as a jury room. Aspiring to wisdom is, or should be, a requirement for every family and every motley amalgamation of would-be citizenry, for wisdom is a form of immortality, a way of giving back to the group (and preserving its survival) long after you have ceased to be a part of it.

But mere hunger for wisdom does not necessarily mean that hunger will be sated. Starvation is a fact of life and a fact of evolution; biologically, our bodies keep track of a very tight clock on the countdown to our next meal, lest we perish, and these needs seem to be hard-wired into the neural machinery of decision making and choice. So why do we hunger so much for something as insubstantial as wisdom?

I think it is because humans, unlike all other creatures on Earth, have that second relentless cognitive clock ticking inside their heads, counting down in a covert yet unassailably certain way the hours and minutes of our remaining time on Earth. Just as we literally hunger for food and water to forestall physiological death, we figuratively hunger for wisdom to forestall spiritual and existential death. As those founding fathers of philosophy, those original "lovers of wisdom," knew only too well, the path of the well-lived, virtuous life has meaning precisely because that path arrives, for every living soul, by whatever circuitous route, at exactly the same destination.

In 2007, *The Last Lecture,* by Randy Pausch, became a publishing sensation. Pausch, a young professor of computer science at Carnegie Mellon University and a self-described "recovering jerk," had been diagnosed with pancreatic cancer in 2006. His last academic lecture was a compendium of the advice, observations, and wisdom about life he wished to leave to his wife, his three young children, his students, and, it turns out, millions of readers throughout the world. No one would confuse Randy

Pausch with Socrates or Montaigne, but we all paused to pay attention. Why? On July 25, 2008, the day he died, I happened to hear a pre-recorded interview with Pausch on the car radio. He said that his terminal diagnosis, and the tragedy of his impending death, gave his final lecture a "moral authority" that it would not otherwise have had.

We don't need the best-seller list—or, for that matter, cognitive neuroscience—to tell us that death focuses the mind marvelously on the meaning of life. In the very first lines of a famous essay entitled "That to Philosophize Is to Learn to Die," Montaigne wrote, "Cicero says that to philosophize is nothing else but to prepare for death. This is because study and contemplation draw our soul out of us to some extent and keep it busy outside the body; which is a sort of apprenticeship and semblance of death. Or else it is because all the wisdom and reasoning in the world boils down finally to this point: to teach us not to be afraid to die." This "drawing out" of the soul, this ancient concept of mindfulness, forces us to rise above the petty urgencies of the everyday, abides a different clock, requires a different focus, demands a different kind of courage. As Julian Barnes noted in his recent meditation about mortality, Montaigne "is where our modern thinking about death begins."

It is no coincidence that the cultural stereotype of the sage has always been the old man or the wizened woman; it is precisely that wrinkled demographic that is actuarially "closer to death," and whose observations and advice thus have the moral authority to command our attention and respect. It is no coincidence that Socrates again and again returned to the liberating effects of death, and to his profound conviction that wisdom flourished in the curious fertilizer of impending mortality, during the thirty days that he contemplated it. Montaigne sounded many variations on the same theme. The composure with which we face the prospect of dying, he wrote, is "without doubt the most noteworthy action of human life." As a Stoic philosopher told the young Roman Tullius Marcellinus as he battled a terminal disease, "It is no great thing to live—your valets and animals live—but it is a great thing to die honorably, wisely, and with constancy."

"Wisely, and with constancy": What kind of death is that? And why does wisdom cohabit so chummily with death? The answer, of course, is that wisdom informs all the decisions we make, momentous and trivial, early in life or late, personal and public, as we strive to conduct our lives in the most meaningful, virtuous, and fulfilling way. We usually don't get around to that self-assessed scorecard on how we've done until, if fortune bestows the luxury of reflection, we are about to lose our lease on life (many, to whom death comes swiftly and unexpectedly, never even have the privilege or burden of conducting that review—all the more reason to keep a constant, running tab).

The notion of constancy is especially important. As I was writing this book—indeed, making notes and contemplating how I would conclude it—my father, after a long illness, died in the summer of 2008. He was one of those "Greatest Generation" guys born in humble small-town circumstances, whose childhood was dominated by the adversity of the Great Depression and whose young adulthood was shaped by combat in World War II (which for him extended well beyond the war, for he was shot by a German sniper a week before hostilities ended in Europe and spent the better part of the next two years in a military hospital). As we reviewed the events and values of his life, as families do when they lose a loved one, I realized that he embodied, in his typically quiet and modest way, many of the values that form the behavioral foundation of wisdom: immense patience, the fearless aggregation of knowledge, a principled compassion not only for humans but for animals, a moral judgment and sense of fairness that must have been cast in a foundry, and an other-centeredness that allowed all his intellectual energies and emotional resources to be poured into the single most important project in his life, his family. We are culturally conditioned to think of wisdom in Lincoln-esque or Solomonic terms, but in terms of day-to-day, lifelong impact, it is hard to imagine a grander achievement on life's stage than to have been a wise parent.

Death jostles the viewfinder through which we look at life most of the time. If we're lucky, it slows the clock of our quotidian frenzies long enough for us to glimpse a more distant future, see a more worthy goal, imagine a better self. This pause, this form of framing, is harder than ever to achieve nowadays, because so many of our modern technologies produce "personal" devices that collapse time and manufacture urgency— faster computers, phones that make us perpetually reachable, twitters of constant thoughts, webs of interaction that vastly increase common knowledge, yet somehow deprive us of that apprenticed learning that leads to wisdom; this digital haze obscures our view of the future and keeps our focus ever more relentlessly on the present, with ever more insistence on speed as a virtue in and of itself. Only the brave, the strong, the mindful, and perhaps the poor (who cannot afford this arsenal of self-absorption) have any chance of resisting the technological pressure to live faster, more in the present, less contemplatively. They at least can legitimately imagine a future that includes their own wisdom.

The earliest philosophers said it all before: Death—its inevitability, its serial visitations to all those around us, ultimately its patient or greedy circling of us—is what sharpens our eye for details about the well-lived life and whets the edges of our hearts and minds as we cleave to decisions and behaviors that aspire to be wise. We do so, and will always want to do so, because wisdom counsels a goodness that extends beyond the membrane of ego and our utile self-interests, radiates outward in an enveloping generative energy that empowers loved ones, kin, students, our various tribes of affiliation, and, if we're particularly lucky and particularly wise, our larger body politic. It gives us a chance to perform the magic of being simultaneously selfless and self-improved.

If, as the scholars who have studied it most intensely maintain, wisdom is essentially an unattainable ideal, it is one of those illusions—like the power of the human mind to imagine a hopeful future or the power of the human heart to surmount a traumatic past—in which even the residue of failure, which may be all we're left with at the end of the day, will leave us better off than if we had not dared to be wise in the first place.

As long as there is a biological moment at which we cease to be, and the unique human capacity to ponder the ramifications of that moment, there will always be a hunger for wisdom. The future of wisdom, in that sense, is boundless and eternal. And if it remains a distant and elusive utopian kind of aspiration, well, it's always good to keep one foot firmly planted in the world of practicality, but when it comes to venturing into the future, be it personal, familial, or collective, I'm with Oscar Wilde, that bighearted cartographer of the possible, when he said, "A map of the world that does not include Utopia is not worth even glancing at."

If only wisdom, too, could be reduced to a spot on a map. If only we had some spiritual form of GPS that would guide us to a particular set of coordinates in a defined physical (or, for that matter, neurophysiological) place, with its own situational landscape, its own emotional climate, its own varying but at least vaguely predictable local customs and dynamics, a destination where we could travel with a reasonable probability of arriving and a reasonable probability of finding there what we would expect to find. But wisdom is too big, too diffuse, and too mysterious for such easy capture; it builds out of what Emerson called "secret currents of might and mind." Even though those currents bear us to a single and inescapable destination in death, just as every river eventually finds the sea, we get there by an infinite number of different paths created by distinctly unique lives.

Ultimately, the route to that other destination we more happily seek, wisdom, may be much more upstream and elusive than we think. We have been acculturated to think of wisdom as a uniquely human virtue, the product of cognition and emotional intelligence and adaptability. But there may be something like wisdom in other creatures. John Meacham found wisdom in cows. Stanley Falkow, the award-winning microbiologist, has described the wisdom of bacteria. Edward O. Wilson has spent a career discovering the wisdom of ants. Vivian Clayton reminded me of precisely this point the last time I met her.

Clayton and I had lunch on a dreary November day in 2007. Although she has not published a single word of research on wisdom in about thirty years, she still feels great affinity for the subject, and almost a matriarchal zeal about the tutelage of wisdom as an idea worth nurturing and protecting and thinking about. As we discussed various aspects of wisdom over, appropriately, the kind of cuisine that would not be unfamiliar to current adherents of the Buddha, she kept trying to tease out my thoughts on various aspects of wisdom. It was Clayton, of course, who had been a minder of bees, and who ascribed a certain wisdom to the synchronous hum of a hive unified and activated by a shared, altruistic mission, and over lunch she kept edging toward a point she wanted to leave me with. It had to do, of course, with bees.

You'd be surprised how often bees buzz through the touchstone texts of Western philosophy. After visiting Clayton in California the first time in 2007, I kept running across references to bees: in Aristotle, of course, but also in Marcus Aurelius and Franklin and Emerson and Darwin. The references often had to do with selflessness, social organization, and perhaps the insect equivalent of gamma oscillation—a synchronizing wave of energy that brought an entire social group into perfect harmony.

It is very hard, Clayton pointed out, to find the queen in a hive. That was her way of saying that we tend to focus so much energy and attention in an attempt to identify wise people that we sometimes lose sight of the larger idea, the *thing* itself. "Oddly enough," she said, "in insect life, there is this wonderful colony that produces this wonderful substance, and it is so well coordinated that scientists kvell about it and say, 'If only humans could get along like this.' And one of the hardest things to do is to find the queen, which is the cause of this whole society." In fact, Clayton had once written a drosh, a kind of lay sermon for her local congregation, on the metaphorical significance of not being able to find the queen.

"To me," she continued, "that signifies in a concrete way that the origin of these very mysterious things is just that—hard to find, but there. This type of knowledge is always going to be a little elusive and evasive, but it's always there. Bees are a carnal example of that."

Taking her cue from the hive, Clayton then spoke of a holistic, harmonic buzz of shared knowledge. "This timeless knowledge—it *pulses*. The wise person can harness it but doesn't own it. This pulse of the universe, this fabric of the universe, this . . . I'm not sure of the metaphor," she continued, fumbling for the words that struggled to capture such an elusive prey.

Finally, she left me with a parting nudge of advice, with which I am happy to comply. "What I'm trying to say," she said, "is, Leave some mystery there."

ACKNOWLEDGMENTS: CONFUCIUS SAYS . . .

As Confucius once said, "The superior man is sparing in words. " But any author would gladly risk Confucian wrath when it comes to acknowledging all the invisible hands that help nurture a book into print.

First, a quick disclaimer as a form of thanks. Anyone who presumes to tackle as weighty a subject as wisdom is asking for cosmic comeuppance—either from the large, discriminating circle of general readers who collectively know more about this rich subject than any one writer can hope to master, or from the smaller (and possibly less forgiving) circle of friends and loved ones who might reasonably wish to have seen a bit more wisdom from the author on a daily basis. To both constituencies, I express gratitude in advance for their understanding of errors, omissions, and inconsistencies. Wisdom—the binding of the current volume notwithstanding—is always a work in progress.

I've always wanted to write a book about neuroscience; I just never expected this to be the one. I attribute this unexpected turn of events to several neuroscientists who primed me for this task even though they had nothing to do with steering me in the direction of wisdom (and, quite possibly, might distance themselves from my final destination). So let me first thank my four neuro-Virgils, who have over the course of two decades guided my way through the increasingly thick forest of brain science.

I first interviewed Eric Kandel of Columbia University in 1985. This is another way of saying that not only was I introduced to the biology of memory more than twenty years ago but have been schooled for many years to think of this quintessential human quality as an everyday process that converts our experiences into neurological changes at the molecular level. In 1997, I was privileged to describe to readers of *The New York Times Magazine* this world of neural proteins and activated genes in the creation of memory ("Manipulating Memory"); the notion of neural plasticity is essential to any modern ideas about wisdom.

In 1998, while working on another story for the *Times* magazine ("The Anatomy of Fear"), I received an advanced tutorial on the biology of fear from Joseph LeDoux of New York University, who has been a modern Magellan circumnavigating every node of the emotional part of the brain. Since then, Joe has been a constant and generous source of information, edification, and, not infrequently, correction. Again, at the level of molecules and neural wiring, I was fortunate to be exposed early on to the powerful idea that emotions unconsciously guide many of our thoughts and actions, and Joe has been an exemplary source in conveying that message.

With the advent of functional magnetic resonance imaging (fMRI), my journalism about neuroscience became more participatory, and I was again lucky to entrust my brain to several leading figures in the field. As part of that same story on the biology of fear, I underwent my first "journalistic" MRI brain scan in the summer of 1998, as a participant in a fear experiment conducted by Elizabeth Phelps, then at Yale University. You could say I passed with flying colors, in the sense that the mere anticipation of receiving a shock made my amygdala (which we science writers often describe as the "center of fear") explode like fireworks designed by the Grucci brothers. Liz, who has since moved to NYU, has consistently lived on the leading edge of social neuroscience; her most recent work shows that our brains can be trained to optimize cognitive performance in a variety of challenging circumstances—hopeful news for anyone interested in wisdom.

Finally, in 1999, I had the good fortune to concoct, with neuroscientist Joy Hirsch, then at Memorial Sloan-Kettering Cancer Center, a series of custom-designed fMRI "experiments" on my brain that appeared in a special millennium issue of *The New York Times Magazine* ("Journey to the Center of My Brain"). "Experiments" is in quotes because, when you study a single brain, you have no grounds to claim any scientific validity. Nonetheless, we did some interesting neural reconnoitering and, I would like to think, broke some conceptual ground that anticipated later experiments that have appeared in the literature. We looked at my brain when it was doing ordinary daily tasks—reacting to cartoons (to explore the human sense of humor), gazing at a photograph of a loved one, and listening to the voices of one's children (to explore the neural basis of emotional memories).

We also wanted to see what was happening under the hood when I was doing the kind of thing I presumably do well, which is to use language and tell stories. So Joy, who now heads the Program for Imaging and Cognitive Sciences at Columbia University, put me in the scanner and then asked me to perform more creative tasks: to make up a story on the fly while viewing a random set of images, to concoct sentences suggested by several pictures, and to invent similes. The MRI images didn't show anything as focused and dramatic as the "center of metaphor" or the "storytelling node" of the brain. But the sheer symphonic sprawl of activation, like those vast lattices of lightning bolts that sometimes spread synaptically across the Western sky, was both thrilling and humbling. There's a lot going on in there, even when we're looking at a Gary Larson cartoon.

I didn't realize it at the time, but it seems to me now that there was a hugely important lesson about wisdom—and neuroscience—in that diffuse sprawl of neural activity. Our modern explorers of the brain are still, like the explorers and cartographers of the sixteenth century, pinning down its fundamental geography. But the action—in wisdom and probably in most other higher-order human thought—is in those neural circuits and connections that are just now coming into focus, not to say the dynamic changes in connective strength in that circuitry that is audited and revised on a daily basis. I've been very fortunate to learn these lessons from masters in the field, and it has been an honor to pass their knowledge along to the general public.

Along with my four neuro-Virgils, I must in the same Florentine spirit thank my estimable patrons at *The New York Times Magazine* for supporting my serial explorations of neuroscientific issues; over the past decade, I've been lucky to have as editors Adam Moss, Jack Rosenthal, Gerald Marzorati, Katherine Bouton, Stephen Dubner, Michael Pollan, and Vera Titunik. Several years ago, more specifically, I received an incredible gift from the magazine when Vera asked me to look into an article about "wisdom research" for a special issue devoted to baby boomers, which sparked my initial interest in the topic. In addition to Vera, who edited the original piece, I am grateful to Robert Wallace, who did some of the initial conceptual spadework for the story; Jim Schachter, who curated the entire issue; and Eric Nash, who backstopped my accuracy.

If part of wisdom is the ability to keep old friends while gaining new ones, I—and this book—have been triply blessed. When I first met Wallace Matson in 1990, on the lawn of Villa Serbelloni in Bellagio, Italy, he confided, "Croquet is my game." I momentarily wondered what kind of person would dare boast about such an arcane skill; he turned out to be a superb classicist and twentieth-century philosopher. From his wonderful translation of Epictetus to his excellent history of philosophy to his sage advice on matters philosophical, Wallace has been immensely helpful (and indulgent) as I stumbled over his turf. Another old and dear friend, Nelson Smith, has steeped himself so thoroughly and thoughtfully in the literature of philosophy that he served as a one-man rump dissertation committee—challenging my thoughts, questioning my assertions, and forcing me to think and write clearly—at our longtime venue for philosophical disputation, Fanelli's Restaurant & Bar. Finally, in reporting on the early empirical research on wisdom, I not only managed to track down one of the seminal figures in Vivian Clayton, but acquired a friend in the process. Her vast knowledge of the history, psychology, and theology of wisdom has been an enormous help, and she has selflessly shared advice, suggestions, thoughtful criticism, droshes, and even the fine points of bee husbandry, all of which have made this a better book.

In addition to the aforementioned three readers, I am also grateful to people who commented on portions of the manuscript, including George Ainslie, Fredda Blanchard-Fields, Roger Blumberg, Richard Davidson, Ernst Fehr, Vittorio Gallese, Ursula Goodenough, James Gross, Donald Hambrick, Rabbi Robert A. Harris, Stephen Kosslyn, Joseph LeDoux, Michael Lotze, Thomas Murray, Thomas O'Neill, Karen Parker, Matthieu Ricard, Jacqui Smith, and Stephen Spear.

As all writers know, the journey to completion is fraught with uncertainty and eased by the kindness of old friends and new acquaintances, who oil the gears of composition with a kind word, a useful suggestion, a provocative thought, a satisfying meal. My colleagues at Columbia University (especially Marguerite Holloway and Bill Grueskin) and at New York University (especially Dan Fagin) have been wonderfully supportive. In the community of science writers, I am grateful for the long-running encouragement of Charles Mann, John Benditt, Gary Taubes, Robin Marantz Henig, Dennis Overbye, David Shenk, Robert Bazell, Amy Harmon, and Michael Pollan. It's also a pleasure to thank Blythe Carey and Liam for good company on the darkest of days; Bob Ray and the late Marion Stocking for never losing interest in an old student; Joel Kaye for his Aristotelian nudges; Thea Lurie for her boundless encouragements; Michael Gazzaniga for neuro-clarifications; A. O. Scott for many helpful discussions, on ice and off; Justine Henning for her rich knowledge of Judaica (a nod to Dahlia Bernstein on this point, too); Joe Helguera for a lifetime of philosophical tips; Richard Klug and Kate Stearns for hospitality and heartening words; Peggy Northrop and Sean Elder for longtime support; Steve Burnett for his wise and whimsical pen-and-ink encouragements; Welker White and Damian Young for their steady support; David Black for reminding me of purposeful goals; Yael Niv for pointing out interesting papers; Josh Greene, who can always be counted on for a stimulating conversation; Alex and Sarah Prud'homme for neighborly commiseration and meals; Ellen Rudolph for unconditional support worthy of a blood relative; Jessica Nicoll and Barry Oreck for Adirondack hospitality at the high point of neuro-stress; and, as always, Joe McElroy for support of every sort in our long-running, pluralistic universe of friendship. I especially thank my dear departed friend Anne Friedberg and Howard Rodman for a decades-long tutorial in courage, intellectual rigor, and personal grace.

This book has had two editorial fathers at Knopf, and both deserve special praise. Marty Asher recognized the science of wisdom as a topic of immense cultural interest, and his enthusiasm and understanding was instrumental in convincing me there was a book here in the first place. His many shrewd suggestions in the conceptualization of the project helped me shape the book. Dan Frank's love of philosophy, rigorous intellect, and (if I may) wise editorial nudges made this a better, sharper, more focused book. The entire production team at Knopf reminded me of what a pleasure it is to interact with people who are exceptionally good at what they do, so I'm delighted to thank: copy editor Carol

Edwards, production editor Ellen Feldman, jacket designer Jason Booher, text designer Robert Olsson, production manager Lisa Montebello, and publicist Kim Thornton. Like Marcus Aurelius's father, editorial assistant Hannah Oberman-Breindel knew when to loosen the reins and when to yank the bit; when Hannah left, Jill Verrillo ably took up the reins and guided us to the finish the line.

Special thanks go to my agent, Melanie Jackson. Our relationship has gone on longer than many marriages (certainly mine) and has many of the best qualities of that kind of union: loyalty, trust, good judgment, unsparing honesty, and mutual respect. She has consistently been this writer's best professional friend for twenty-five years, but her help and perspicacity on this book were spectacular. I can't thank her enough for her discerning and shrewd advice.

As always, the support of family—my parents, Robert and Delores, and my brother, Eric, and his family, as well as my "in-law" family, Gerri and Hindley Mendelsohn, and Susan and Jedd Levine—has been unconditional, unstinting, and steady; if you consider the Italian and Pennsylvania branches of my extended family (as one must do when thinking about the roots of temperament, tradition, ritual, and what we hold to be important), I have been boosted by many, many loving hands. Robert Hall did not live to see the completion of this book, but I'd like to think that his demeanor and decency and courage inflect every page and every thought. He exemplified what Petrarch called "active humility," and I hope this apple hasn't fallen too far from that tree.

Finally, the inner circle. There is no greater test of wisdom on a daily basis than constantly tuning the string of your relationship with loved ones. I am blessed, on a daily basis, with two children, Micaela and Alessandro, who are curious, creative, talk back in all the right ways, and exhibit the patience of an adult (most of the time) when Dad is sequestered in his office; there is no greater gift to a parent than the candor of his children, which often induces medicinal doses of humility. But in terms of blessings, I am especially blessed by the love and support (not to mention keen mind and sharp editorial pen) of my lifelong love, Mindy. Wisdom is shaped by the physics of the company you keep, and she, more than anyone I've ever known, has always understood what is most important. That, on a daily basis, makes everything else possible.

NOTES

CHAPTER ONE: WHAT IS WISDOM?

3   On a beautiful fall morning: This account is a distillation of Hall 2001.

7   the Way of life: Jaspers 1962, pp. 49–50.

8   "Some of the wisest and most devout men": Montaigne 2003, p. 332.

8   we know it when we see it: Gewirtz 1996.

8   "fuzzy zone": Baltes 2004, p. 23.

9   article for *The New York Times Magazine:* Hall 2007.

9   definition of wisdom: Sternberg 1990; see also Birren and Svensson 2005, p. 15. The psychologists offering a definition of wisdom in Sternberg's 1990 anthology include: Mihaly Csikszentmihalyi and Kevin Rathunde, who spoke of wisdom "as a cognitive process, as a guide to action, and as an intrinsic reward" (p. 48); Gisela Labouvie-Vief, who described a "smooth and relatively balanced dialogue" between "experiential richness and fluidity" on the one hand and "logical cohesion and stability" on the other (p. 53); Patricia Kennedy Arlin, who observed that wisdom "is more a matter of interrogatives rather than of declaratives" (p.230); and John A. Meacham, who argued that "the essence of wisdom is to hold the attitude that knowledge is fallible and to strive for a balance between knowing and doubting (p. 181)."

9   "To understand wisdom fully": Sternberg 1990, p. 3.

10 "We ought to seek out virtue": Plutarch 1960, p. 165.

11 psychologists in Canada conducted a study: Paulhus et al. 2002.

12 Pythagoras—who gave us the word *philosophy:* Matson 1987a, pp. 18–21.

13 even the closest friends of Jesus Christ: Jaspers 1962, p. 71.

13 As a recent art exhibit: "Worshiping Women: Ritual and Reality in Classical Athens," Onassis Cultural Center, New York, 2008. See also Murphy 1998 and Buchmann and Spiegel 1994. Murphy, in describing the absence of "literate" women in the Bible, traces the emergence of an iconic "wise woman" (pp. 86–93, 101–3).

13 "Birth and death": Cotter 2008.

14 "The most manifest sign of wisdom": Montaigne quoted in Csikszentmihalyi and Rathunde 1990, p. 37; for recent neuroscientific support, see Sharot et al. 2007.

14 "There have been as great Souls": Franklin 1997, p. 1192.

14 "about people working out the interior problems": Michaelas 2008, p. 245.

15 a meeting at New York University: "Neuroeconomics: Decision Making and the Brain," New York University, January 11–13, 2008.

15 As countless popular science books: For recent examples of the genre, see Ariely 2008;

Gladwell 2005; Lehrer 2009; Montague 2006; Thaler and Sunstein 2008. For basic brain science, I have used Carter 1998; Gazzaniga 2008; Pinker 2002; and Posner and Raichle 1997.

15 "Decision theory works very well": Vernon Smith, "Experimental Economics and Neuroeconomics," New York, January 12, 2008.

15 "hyperbolic discount curve": David Laibson, Harvard University, "Multiple Systems Hypothesis of Brain Organization," New York, January 12, 2008.

17 a $2 million research program called "Defining Wisdom": For more information on the project, see www.wisdomresearch.org; information on grant process and award from John Cacioppo, director, Center for Cognitive and Social Neuroscience, University of Chicago, telephone interview, August 20, 2008, and from "2008 Defining Wisdom Grant Competition: Award Announcements," press release, Arete Initiative, September 3, 2008. See also Cacioppo, Visser, and Pickett 2006.

17 "Wisdom presumably has something to do with memory": Stephen Kosslyn, telephone interview, August 4, 2008.

17 "framing": For information on "framing," see Kahneman 2002. For a more explicit connection between wisdom and neuroscience, see Meeks and Jeste 2009.

19 science represents that rare balance: Medawar 1984, pp. 46, 53.

19 "a topic at the interface": Baltes 2004, p. 5.

CHAPTER TWO: THE WISEST MAN IN THE WORLD

20 Chaerephon made his way: The Chaerephon anecdote was related by Socrates at his trial; see Plato 2003, pp. 43–44.

20 "this false notoriety": Ibid., p. 43.

21 There is divine wisdom: For an early history of wisdom and its various ancient manifestations, see Robinson 1990; Birren and Svensson 2005; Curnow 1999, pp. 10–117.

21 "cycle of labors": Plato 2003, p. 45.

21 "I am only too conscious": Ibid., p. 44.

21 "a man with a high reputation for wisdom": Ibid.

21 "although in many people's opinion": Ibid.

22 "thinks that he knows something": Ibid.

22 "comparative ignoramus": Ibid., p. 45.

22 "could have explained those poems better": Ibid.

22 "on the strength of their technical proficiency": Ibid, p. 46.

22 "I made myself spokesman": Ibid.

23 Socrates gave flesh: For details of Socrates' appearance and life, see Jaspers 1962, pp. 5–6. In Jaspers's rendition (p. 5), "He was an ugly man, with bulging eyes. His stub nose, thick lips, big belly, and squat build suggest a Silenus or satyr."

23 "stubborn perversity": Plato 2003, p. 65.

23 "Axial Age": Jaspers 2003, p. 99.

24 "No man on earth": Plato 2003, p. 58.

25 "Their concern . . . was not mere knowledge": Jaspers 1962, p. 93.

25 The Buddha, whose penniless wandering: Armstrong 2001, pp. 135–39; Wallis 2007, p. xii.

25 For the last ten years of his life: Strathern 1999, pp. 42–43.

25 "Because Socrates alone": Montaigne 2003, p. 323.

26 Some historians of philosophy argue: Matson 1987a, pp. 6–7; Robinson 1990, pp. 13–14.

26 "mutation" in thought: Matson 1987a, pp. 15–17.

26 "Homer and Hesiod": Ibid., pp. 6–7.

27 "The river / where you set": Heraclitus 2001, p. 27.

27 During a rare period of relative peace: Pomeroy et al. 1999, pp. 215–45.

27 "belong to a city": Plato 2003, p. 56.

28 "We Athenians . . . are able to judge": Matson 1987a, p. 66.

28 "the body fills us with loves and desires": Plato 2003, p. 127.

29 It is also difficult to read these lines: For the neuroscientific notion of embodiedness, see Dolan 2006.

29 "Will a true lover of wisdom": Plato 2003, p. 129.

29 "Wisdom diverged from the knowable": Birren and Svensson 2005, p. 8.

29 Born in the sixth century B.C.: This account of the life of Confucius is based on Confucius 2000, pp. xi–xxii, 7–11; Jaspers 1962, pp. 41–63; Strathern 1999.

30 "When I was thirty": Confucius, quoted in Strathern 1999, p. 13.

30 "This grounding in everyday horrors": Strathern 1999, p. 15.

30 Most important is goodness (or *gen*): For the concept of *gen* (often rendered also as *jen*), I have drawn upon Confucius 2000, pp. 21–23; Jaspers 1962, pp. 49–50.

31 "Not to act when justice commands": Confucius, quoted in Strathern 1999, p. 16.

31 "For Confucius, morality was all about *involvement*": Strathern 1999, p. 24.

31 He was said to have been hugely successful: Ibid., p. 41.

31 "unusual clothing": Ibid.

31 "The great mountain must crumble": Confucius, quoted in Strathern 1999, p. 48.

32 "To know what you know": Confucius 2000, p. 83.

32 "When Prince Siddhartha Gautama abandoned: For this account of the Buddha's life, I have relied primarily on Strong 2001 and 2002; Armstrong 2001; Wallis 2007.

32 "adamantine throne"; "I will not uncross my legs": Strong 2001, p. 17.

32 "backslider": Wallis 2007, p. xxxiii.

32 "living luxuriously": Ibid.

32 "Wisdom lies in stilling all desire": Birren and Svensson 2005, p. 8.

32 "strives to annul": Jaspers 1962, p. 93.

33 "Conquerors are those who": Wallis 2007, p. xxxii.

33 "middle way": Ibid., p. 36.

33 "complete attentiveness": Ibid., p. 44.

33 "And there arose in me": Ibid.

33 "censured by the wise": Ibid., p. 23.

33 "the superlative practitioner": Ibid., p 25.

34 Saint Augustine distinguished: For the distinction between *sapientia* and *scientia,* Birren and Svensson 2005, p. 7; see Rice 1958, pp. 1–3.

35 "a philanthropic catalyst for discovery": Templeton Foundation Web site: http://www .templeton.org. For views of religious agenda, see Horgan 2006; Carroll 2008, and Dawkins 2006, pp. 151–54. (Horgan quotes physicist Sean Carroll as saying the foundation "seeks to blur the line between science and religion"; also viewable at http://www .edge.org/3rd_culture/horgan06/horgan06_index.html).

35 "intellectual enemy" of wisdom: Baltes 2004, pp. 56, 75.

36 "Augustine tied *sapientia* and Christianity together": Rice 1958, p. 3.

37 "The Idea of wisdom": Kant, quoted in Birren and Svensson 2005, p. 11.

37 "where knowledge is possible": Ibid.; Robinson 1990, p. 22.

37 "did not possess wisdom": Birren and Svensson 2005, p. 11.

37 "the final appearance of the idea": Revel and Ricard 1999, p. 215.

37 "Wisdom . . . is an understanding": Nozick 1989, p. 267.

38 A decisive turn in the formal history of wisdom research: Information on the White
    House conference taken from Goleman 1988; Erikson, 1959, p. 2. The graph was initially
    published in Erikson, p. 166.

CHAPTER THREE: HEART AND MIND

39 "My father was forty-one": This and other biographical details are from Vivian Clayton,
    personal interview, Orinda, California, March 27, 2007.

40 She is generally recognized as the first psychologist to ask: Personal interviews with
    Robert Sternberg (2007), Jacqui Smith (2007), and Monika Ardelt (2007); see also Stern-
    berg 1985, p. 609; Baltes 2004, pp. 194–95; Ardelt 2005a, pp. xi–xii.

40 "on the first great longitudinal studies": Vaillant 1977, p. 201.

41 Erikson viewed wisdom as a central feature: The idea of wisdom appears in at least three
    of Erikson's early books, but with little definitional or methodological elaboration. In
    Identity and the Life Cycle (1959), he recounted the background of his contribution to the
    1950 White House conference on childhood development; a work sheet in an appendix
    includes a graph showing the eight stages of development, with the eighth stage charac-
    terized by "Wisdom" (p. 166). In *Identity: Youth and Crisis* (1968), he expanded on wis-
    dom in general terms, with the remarks quoted in the main text ("meaningful old age"
    and "Strength here takes," pp. 140–41; I am grateful to *New York Times* reader Tom
    Frangicetto for bringing this later version to my attention). Finally, in later editions of
    Childhood and Society (1950), Erikson also appended a footnote to the "Eight Ages of
    Man" chart, in which he describes wisdom as a "basic virtue" in the final psychosocial
    stage of life (pp. 273–74). Erikson, of course, addressed the issue of wisdom in his later
    work; I am interested here in early suggestions that it would be a topic, as he himself
    stated, worthy of methodological exploration.

42 Half a century ago, gerontologists dominated: Information on gerontological bias is from
    Laura Carstensen, personal interview, Stanford, California, March 26, 2007.

42 One of the leading voices: For Birren's role in pushing positive aspects of aging and the
    general shift in gerontological psychology, I have relied on multiple interviews with
    Vivian Clayton and Laura Carstensen. See also Sternberg 1985.

42 Then a psychologist at Pennsylvania State University: Baltes 1999, pp. 7–26.

43 "to think logically, to conceptualize": This and subsequent quotes are from Clayton 1982.

43 *emotional* intelligence: Goleman 1995.

43 The famous story about the wisdom of King Solomon: The basic account of Solomon's
    life is in 1 Kings; I have used the *New Oxford Annotated Bible*. On scholars raising ques-
    tions about Solomon's wisdom, see Baltes 2004, pp. 58–59.

44 Hebrew narratives: For the Hebrew tradition of wisdom, I have consulted Alter 2004; the
    New Oxford Annotated Bible; Telushkin 1994; Gribetz 1997; Wiesel 2003; *The Jewish
    Study Bible* 2004.

44 "lust to the eyes": Alter 2004, p. 24.

44 "to make one wise": Ibid., p. 25.

44 "For God knows": Ibid.

44 Later on in Genesis: On Abraham and the "equation" of age and wisdom, see *Etz Hayim*
    2001, pp. 130–31.

45 "What emerged from that analysis": Vivian Clayton, personal interview.

46 "Our decision to sample lawyers": This and other details from the 1974 study are from
    Vivian Clayton, "Dear Participant" letter, December 20, 1974, private communication to
    the author.

47 Clayton published several groundbreaking papers: Clayton 1975; Clayton and Birren

1980; Clayton 1982. Her dissertation was summarized in Clayton and Schaie 1979, unpublished ms., courtesy of the author.

47 "The function of intelligence": Clayton 1982, p. 315.

47 "a big deal": Robert Sternberg, telephone interview, March 15, 2007. See also Sternberg 1985, p. 609.

47 "seminal work that triggered": Jacqui Smith, telephone interview, March 23, 2007.

48 "I'd just stand at the juncture": Clayton 1982, p. 320. This anecdote was originally reported by the Baltes group, but in a study of elderly patients that had nothing explicitly to do with wisdom.

48 Clayton explicity made the point: Ibid., p. 316.

48 "A 15-year-old girl wants to get married": Baltes and Staudinger 2000, p. 136. Other examples of similar "wisdom dilemmas" were kindly provided by Jacqui Smith, personal communication, May 2007.

49 "bring research on wisdom into the laboratory": Baltes and Smith 2008, p. 58.

49 created the intellectual space for experimentation: Clayton 1982; Sternberg 1985.

49 "spurred by a motivation to identify": Baltes and Staudinger 2000, p. 122.

49 Baltes wrote and posted on the Internet: Baltes 2004 (reviewed in Smith 2007).

49 "an expert knowledge system": Baltes and Staudinger 2000, p. 122.

50 "wisdom is a collectively anchored product": Ibid., p. 130.

50 Baltes wrote a brief history of proverbs: Baltes 2004, pp. 12–15.

50 "Marrying several wives is human": Ibid., p. 46.

50 "He who spurns his father's discipline": Ibid., p. 60.

50 As Baltes pointed out, many famous maxims: Ibid., pp. 12–13. Baltes noted, for example, that "Clothes make the man" clearly contradicts "Don't judge a book by its cover" (p. 13).

51 "wisdom deals with difficult problems": Ibid. p. 22.

52 "could serve as a prototype": Ibid., p. 32.

52 "Probably only somebody who has tenure": Jacqui Smith, telephone interview.

52 "moved the field forward": Robert Sternberg, telephone interview.

52 "It's great work": Laura Carstensen, personal interview. Later on, the Berlin group emphasized more of a role for emotion (Baltes and Kunzmann 2004).

52 "translated some of the components of wisdom": Baltes and Smith 2008, p. 58.

53 "there are no correct answers": Ibid., p. 59.

53 Consider another profile in wisdom: Details of "Claire" 's life, including quotes, come from Ardelt 2005b, pp. 11–17; "Claire," telephone interview, March 15, 2007.

53 "Three-Dimensional Wisdom Scale": Ardelt 2005b, pp. 8–9; Ardelt 2000.

53 "You can't just ask people": Monika Ardelt, telephone interview, March 7, 2007.

54 In 1997, Ardelt and her colleagues received a grant: Ardelt 2005b, p. 7.

54 "Successfully coping with crises and hardships": Ibid., p. 7.

55 "how relatively wise older people cope with life crises": Ibid., pp. 7–8.

55 "James," for example: Details of James's life, including quotes, come from Ardelt 2005b, pp. 11–17; "James," telephone interview, April 26, 2007.

55 "I don't think we have a clear definition yet": Monika Ardelt, telephone interview. For a flavor of the definitional disagreement, see Ardelt 2004.

56 "I reached a fork in the road": Vivian Clayton, personal interview.

57 "hive psychology": Haidt, Seder, and Kesebir 2008.

CHAPTER FOUR: EMOTIONAL REGULATION

61 endless series of cognitive, behavioral, and emotional assessments: The description of these assessments and remarks concerning them are taken from Nancy Lynne Schmitt,

personal interview, Stanford, California, March 26, 2007; N. L. Schmitt, August 11, 2008, personal communication.

62 "doing what husbands and wives are supposed to do": Jan Post, Santa Rosa, Calif., telephone interview, March 23, 2007.

62 "Happiness is a seven": This remark related to emotions on scale of one to seven.

63 "To be evenminded": Heraclitus 2001, p. 71.

63 "He that can compose himself": Franklin 1997, p. 1205.

63 What the Stanford researchers have found: For original "beeper experiment" results, see Carstensen et al. 2000. Preliminary results from the 2007 data-gathering period, unpublished at the time this was written, were "quite consistent" with the earlier findings (Laura Carstensen, telephone interview, September 9, 2008).

63 "According to our theory": This and other quotes from Laura Carstensen, personal interview, Stanford, California, March 26, 2007. For a good, lay-friendly account of socioemotional selectivity theory, see Carstensen 2007; for a more scientific take, see Cartensen 2006.

64 Many researchers are familiar with a classic case study: For the story of Job, I have used the *New Oxford Annotated Bible* version.

64 the story of Job is usually packaged: For "Job the Impatient," see Ginsberg 1967.

64 "Why should I not be impatient?": Job 21:4.

64 "It is not the old": Job 32:9.

64 "I hold fast my righteousness": Job 27:6.

65 "Surely vexation kills the fool": Job 5:2.

65 In his groundbreaking 1884 essay: James 1884.

65 A dozen years earlier, Darwin had tackled: Darwin 1955.

65 Coming from the other direction: For recent books on the neurobiology of emotion, see Damasio 1994; LeDoux 1996; Haidt 2006.

65 "Consciousness is the weird thing": LeDoux, quoted in Hall 1999a, p. 47.

66 But it has been only in the last decade: "Reappraisal" and other aspects of emotion regulation are discussed in Gross 2008, pp. 497–512. See also McRae et al. 2008; Ochsner and Gross 2008.

66 "crucial" issue: Gross 2008, p. 509.

66 In one recent study: Goldin et al. 2008.

66 "There's an emerging understanding": James Gross, Department of Psychology, Stanford University, telephone interview, June 11, 2009.

67 Laura Carstensen's professional epiphany: Laura Carstensen, personal interview.

68 "As we approach old age": Baltes 2004, p. 27.

68 "Mortality is only one of the commonplaces": Kekes 1983, p. 282.

68 Perhaps the most surprising scientific destination: On differences between old and young, see Mather et al. 2004; Carstensen 2006.

71 Carstensen and her colleagues have actually seen: Samanez-Larkin et al. 2008; Nielsen, Knutson, and Carstensen 2008.

71 This live-for-the-moment, "carpe diem" effect: Fung and Carstensen 2006.

72 Young people tend to be steep temporal discounters: Ainslie 2001, p. 34.

72 "There's no doubt": Paul Glimcher, personal interview, New York, July 2, 2008.

72 a recent experiment by Elizabeth A. Phelps: Sokol-Hessner et al. 2009.

72 "The really important decisions": Lisbeth Nielsen, telephone interview, September 5, 2008.

73 "I do indeed lose my temper": Montaigne 2003, p. 660.

73　Fredda Blanchard-Fields of the Georgia Institute of Technology: Blanchard-Fields, Mienaltowski, and Seay 2007.

73　Davidson and his colleagues looked at patterns of brain activity: Urry et al. 2006.

74　"Those people who are good at regulating": Richard J. Davidson, telephone interview, March 15, 2007.

74　Several years ago, Carstensen and her colleagues: Mather et al. 2004.

75　men expend less cognitive effort: McRae et al. 2008.

75　The experiment investigated how emotion colors: Sharot et al. 2007. See also, in the same issue of *Nature,* "Making the Paper: Tali Sharot," p. xiii.

75　Phelps says that the optimism experiment: Elizabeth Phelps, personal interviews, New York, November 14, 2007, and August 20, 2008.

76　"not constrained by reality": Sharot et al. 2007, p. 104.

77　"Expecting positive events": Ibid.

77　Laura Carstensen and Corinna Lockenhoff: Carstensen and Lockenhoff 2003.

77　"Information processing capacity": Laura Carstensen, personal interview.

77　"Individuals having any advantage": Darwin 2003, p. 600.

78　According to the well-known "grandparent hypothesis": Hawkes 2004.

78　"evolutionary selection should have *favored* skills": Carstensen and Lockenhoff 2003, p. 155.

CHAPTER FIVE: KNOWING WHAT'S IMPORTANT

79　Glimcher claims no expertise in wisdom: All quotes, unless otherwise noted, are from Paul Glimcher, personal interviews, New York, July 2, 2008, and August 28, 2008.

81　"In its power to overturn": Churchland 1986, p. 481.

82　In the fall of 2006 and the spring of 2007: For the Glimcher lab's recent work on decision making and temporal discounting, see Kable and Glimcher 2007; Glimcher, Kable, and Louie 2007.

82　for more about MRI: A note on the use of magnetic resonance imaging (MRI) in brain studies: The scientific rationale for using functional MRI (fMRI) studies is that the imaging machines, through a complicated technology combining the magnetic qualities of atoms inside organic molecules and sophisticated computer programs that process the data, identify regions of the brain where blood flow increases during the performance of experimental tasks (the technical term for this is the "blood oxygen level dependent," or BOLD, signal). The basic idea is that cells put to work by a cognitive task require more energy, and the delivery of the raw materials for this heightened cellular metabolism— oxygen in the blood—can be detected as increased blood flow in certain regions. The level of resolution in most machines can distinguish differential activity in a bit of brain tissue as small as a grain of rice.

There are certain inherent limitations to the technique, and results must be interpreted with care. MRI is essentially *correlative;* it identifies a correlation between behavior and activity in a part of the brain, but not causation, and hence the design of an fMRI study is very important. Functional MRI is also slow and can't resolve activity briefer than about four seconds; electrical recordings of brain activity, by contrast, offer ten times better temporal resolution. Moreover, the heavy reliance on sophisticated computer programs for interpretation of the data has led to criticisms of flawed analysis. Social neuroscientists in particular have been criticized for finding "voodoo correlations" in their fMRI analyses. For background on this debate, see, for example, Abbott 2009; Lakoff 2008, pp. 195–96.

83 the experimenters actually called it "idiosyncratic": Kable and Glimcher 2007, p. 1625.

83 The road to understanding how the brain weighs options: For an overview of reinforcement learning and the latest neuroscience model of valuation and decision making, see Glimcher 2008; Daw, Niv, and Dayan 2005; Dayan 2008; Montague 2006. For a scientific overview of the dopamine system, see Montague, Hyman, and Cohen 2004; for a more lay-friendly recent account of the dopamine reward system, see Lehrer 2008.

84 It involves fishing for crabs: Robb Rutledge, personal communication, June 24, 2009.

86 Bees make decisions: Montague et al 2002; Shafir et al. 2008.

86 a recent article in the journal *Science*: Yang et al. 2008.

87 Peter Dayan's group at University College London: Daw, Niv, and Dayan 2005.

91 A large body of work suggests: There has been a huge amount of research on the role of the unconscious in decision making. Two fascinating recent perspectives on this are Dijksterhuis et al. 2006; Galdi, Arcuri, and Gawronski 2008. The Dijksterhuis paper argues (p. 1005), "Contrary to conventional wisdom, it is not always advantageous to engage in thorough conscious deliberation before choosing."

91 John-Dylan Haynes and his colleagues: Soon et al. 2008; Haynes et al. 2007; Haynes and Rees 2006.

91 "You always have to know something about the domain": John-Dylan Haynes, personal communication, June 29, 2008.

91 "People tend to engage": Dijksterhuis and Nordgren 2006, p. 108.

92 "thought without attention": Ibid., p. 96.

92 "steady attention": Rosen 2008, p. 109.

92 "The faculty of voluntarily bringing back": James 1981, p. 401.

92 "attentional blink": For meditation and attentional blink, see Slagter et al. 2007; for a study on a different aspect of attention and task performance, see Hedden and Gabrieli 2006.

92 "cultivate awareness of stimuli": Jones 2007, p. 497.

93 In a literally delicious set of experiments: Iyengar and Lepper 2000.

94 Neurologists have described an unusual state of paralysis: Frank et al. 2007.

97 "The problem of choice": Montague 2008.

CHAPTER SIX: MORAL REASONING

98 "the eyes of the two were opened up": Alter 2004, p. 24.

98 "From the tree I commanded you": Ibid., p. 25.

99 "Snakes were a symbol": *New Oxford Annotated Bible*, p. 14.

99 "you will become as gods": Alter 2004, pp. 24.

99 chronicled at least as early as the sixth century B.C.: *New Oxford Annotated Bible*, pp. 9–10.

100 "Not to act when justice commands": Confucius, quoted in Strathern 1999, p. 16.

100 "Set your heart on doing good": Buddha, quoted in Haidt 2006, p. 155.

100 "impossible to be practically wise": Aristotle 1908, p. 73.

100 "moral judgments are mediated": Hauser 2006, p. 2.

100 this moral grammar is innate: A note on the neural "grammar" of moral judgment: John Mikhail worked with MIT linguist Noam Chomsky as a graduate student in philosophy, which is when he began to wonder if there was a universal moral grammar in humans, similar to Chomsky's theory about an innate human capacity for language. Mikhail's preliminary findings, extended by Hauser's group at Harvard, suggest that people from many different cultures respond to standard moral dilemmas in a similar way—in other words, they speak the same moral language. Jonathan Haidt's research, by contrast, suggests a strong cultural component to moral reasoning.

100   "Our moral instincts are immune": Hauser 2006, p. xviii.

101   "punishes the cheaters": Jean-Jacques Hublin, Max Planck Institute for Human Evolution, Leipzig, Germany, personal interview, Grenoble, France, October 7, 2007.

101   has begun to attract the attention: See Wright 1994 and Gazzaniga 2005.

101   "As the sensation of disgust primarily arises": Darwin 1955, p. 257.

101   More than a century later: Rozin, Haidt, and Fincher 2009; Chapman et al. 2009.

101   "Of all the differences": Darwin 2004, p. 120.

102   "It is the most noble": Ibid.

102   "ethical intuitionism": Haidt 2001, p. 814.

102   "Moral intuition": Ibid. Haidt's theory of moral judgment is also spelled out in considerable detail in Haidt 2007; Haidt 2006.

102   In an instantly classic (and deeply subversive) experiment: Schnall et al. 2008.

102   An earlier experiment by Haidt: Wheatley and Haidt 2005. For an up-to-date general discussion of Haidt's experiments on visceral disgust influencing moral judgment and where it fits into current research, see Miller 2008a; Pinker 2008.

103   In yet another Haidt experiment: Haidt 2001.

103   It can hardly be a coincidence: Kass 1997; Kass 1999.

104   "Repugnance may be the only voice left": Kass 1997, p. 20.

104   "There are, of course, good reasons": Pinker 2008, p. 58.

104   "Morals excite passions": Hume 1739/1978, p. 457.

105   "From Neural 'Is' to Moral 'Ought'": Greene 2003.

105   Greene set out to study business: Biographical background from Joshua D. Greene, Department of Psychology, Harvard University, Cambridge, Massachusetts, telephone interviews, December 10, 2007, and August 7, 2008. A scientific autobiography is also available at http://www.wjh.harvard.edu/~jgreene/.

105   "I view science": Greene 2003, p. 847.

106   "The Terrible, Horrible": A copy of Greene's doctoral dissertation (for the Department of Philosophy at Princeton University) can be downloaded from the Web site noted above.

106   The "Trolley Problem": Even Wikipedia has a separate entry on the Trolley Problem, describing its early history. According to Greene, Foot introduced the thought experiment in the 1970s in discussing the ethics of abortion, but Thomson "really ran with it."

106   The general consensus: Hauser et al. 2007; Cushman, Young, and Hauser 2006; Young et al. 2007.

106   Hauser and his Harvard colleague: Fiery Cushman, Department of Psychology, Harvard University, Cambridge, Massachusetts, telephone interview, March 17, 2009. For results of these surveys, see Hauser et al. 2007.

107   standing next to a "large": Greene et al. 2001, p. 2105.

107   What they discovered is that moral decisions: Greene et al. 2001; Greene et al. 2004.

107   "up close and personal": Joshua Greene, telephone interview; he has also used this phrase in some of his papers.

108   "The philosophical commentary on these cases": Appiah 2008, p. 91.

108   This sample may represent: In an interview, Greene said the findings remain robust over all age groups.

109   As Greene has pointed out: Greene 2009.

109   On one side are the utilitarians: Ibid., pp. 37–40.

109   On the other side are the so-called deontologists: Ibid.

109   As Kant's foundational work: Kant 1996 and Kant 1999.

109   "We propose that the tension": Greene et al. 2004, p. 398.

110   "moral emergency": Appiah 2008, pp. 96–98.

110　"Holding onto revulsion": Ibid., p. 98.

111　"psychologically speaking": Greene et al. 2004, p. 398.

111　Because this is frontier science: Miller 2008a.

111　people who have suffered injuries to a different part of the brain: On damage to the ventromedial prefrontal cortex, see Koenigs et al. 2007; Talmi and Frith 2007; Moll and de Oliveira-Souza 2007; Moll et al. 2005; Moll et al. 2007.

112　And in a recent study in *Science*: Hsu, Anen, and Quartz 2008.

112　It will take at least ten years: Walter Sinnott-Armstrong; Department of Philosophy, Dartmouth College, Hanover, New Hampshire, telephone interview, March 17, 2009.

112　having "special" emotional relationships: Thomas H. Murray, president, Hastings Center, Garrison, New York, personal interview, New York, April 21, 2009.

113　There were, as he suggests: Greene 2009, p. 43.

113　Jonathan Haidt often uses the metaphor of the elephant: Haidt 2006, p. xi.

113　"People who claim emotion": Jonathan Haidt, telephone interview, March 24, 2009.

CHAPTER SEVEN: COMPASSION

115　laid siege to the town of Weinsberg: The account of the siege and Montaigne's comments about it are from Montaigne 2003, pp. 3–4.

116　Many traditional working definitions of wisdom: Clayton 1982; Sternberg 1990; Sternberg and Jordan 2005.

116　"Never do to others": Confucius 2000, p. 187; according to Luke 6:30, Jesus described the Golden Rule when addressing the crowd in Judea immediately after he named the twelve disciples.

116　"we are hardwired to follow": Pfaff 2007, p. 4. Pfaff's recent book-length account of the biology of the Golden Rule is a particularly good source of background on the neurobiology of these pro-social traits.

117　Indeed, a chilly draft of dispassion blows: Plato 2003, p. 119.

117　"icy": Nietzsche quoted in Stone 1988, p. 146.

117　"crying hysterically": Plato 2003, p. 119.

117　the cries of a "distressed woman": Lutz et al. 2008, p. 2.

117　"the world that comes into view": Davidson and Harrington 2002, p. 20.

118　"Knowledge without compassion": Ibid., p. 18.

118　These researchers have recruited: For the neuroscience of cabdrivers, see Spiers and Maguire 2008; for piano players, see Furuya and Kinoshita 2008.

118　At exactly 8:48 a.m.: These and other descriptions and observations are based on the author's visit to the Waisman Center of Neuroscience, University of Wisconsin, Madison, June 3, 2008.

119　"We have found a correlation": Antoine Lutz, personal interview, Madison, Wisconsin, June 3, 2008.

119　That study reported that expert meditators: Lutz et al. 2004.

120　turn it on "like that": R. J. Davidson, "Order and Disorder in the Emotional Brain," Social Neuroscience Colloquium, Department of Psychology, New York University, March 13, 2008.

120　But since dedicating his life: Matthieu Ricard, personal interview, Madison, Wisconsin, June 3, 2008. Also see Ricard 2006, pp. 4–5.

120　"With a few practitioners": Andy Francis, personal interview, Madison, Wisconsin, June 3, 2008.

121　"It doesn't require any fancy computer interpretation": Richard J. Davidson, personal interview, Madison, Wisconsin, June 3, 2008.

121 A study by the Wisconsin group: Lutz et al. 2008.

122 As an example, he cited a hypothetical case: Matthieu Ricard, personal interview; Matthieu Ricard, personal communication, March 17, 2009.

123 "altered states of consciousness": Davidson and Harrington 2002, p. 107. Davidson's fellow graduate students at the time were Daniel Goleman and Jon Kabat-Zinn (Flanagan 2007, p. 159).

123 Davidson admits that some youthful dabbling: Richard J. Davidson, telephone interview, February 18, 2008.

123 "an acute lack of experiential learning": Davidson and Harrington 2002, p. 107.

123 the Tibetan word for compassion: Ibid., p. 98. Ricard suggests that an alternative meaning for *tsewa* is "tenderness" (Matthieu Ricard, personal communication, June 6, 2008).

123 "In a way," the Dalai Lama has said: Davidson and Harrington 2002, p. 98.

123 "generating compassion": Richard J. Davidson, personal interview.

123 "how we can approach the construct of compassion": Davidson and Harrington 2002, p. 108.

123 "Unconditional Love and Compassion": This and the other distinct forms of Buddhist compassion meditation were noted by Matthieu Ricard, personal communication, June 5, 2008.

124 "These are the Olympic athletes": Hall 2003, p. 46.

124 "unconditional loving-kindness and compassion": Lutz et al. 2004, p. 16,369.

124 "Just as we have done": Richard J. Davidson, personal interview.

125 "The perspective taking": Ibid.

125 A growing number of studies: Lutz et al. 2008; Slagter et al. 2007; Delgado, Gillis, and Phelps 2008; Schiller et al. 2008; Sokol-Hessner et al. 2009. See also Begley 2007.

126 he, too, has published peer-reviewed studies: Kabat-Zinn 1990.

126 can even enhance the performance: Davidson et al. 2003.

126 psychotherapists like Zindel V. Segal: Corcoran and Segal 2008.

126 "Where does that seed come from?": Davidson and Harrington 2002, p. 99.

126 a unique constellation of cells and structures: Iacoboni 2008.

126 "a very strange class of neurons": This and all other quotes by Gallese are from his talk at the "Creativity and the New Biology of Mind" symposium at Columbia University, New York, March 24, 2006.

126 In an oft-told tale: For this account of the discovery of "mirror neurons," I have relied on Iacoboni 2008, pp. 7–11; Gallese, Columbia symposium talk.

127 "These neurons did not only fire": Gallese, Columbia symposium talk.

127 Think about the implications: For recent implications in social neuroscience, see Gallese 2007; Gallese et al. 2009; Rochat et al. 2008; Caggiano et al. 2009.

127 Researchers have learned: Gallese, Columbia symposium talk.

128 "All these areas have been shown to be activated": Ibid.

128 "Quite simply, I believe this work will force us": Iacoboni 2008, p. 7.

128 Raymond Dolan of University College London: Raymond Dolan's talk at the Columbia symposium. For the Conroy passage, see Joyce 1976, pp. 233–36.

128 In a 2004 experiment: Singer et al. 2004.

128 "Such awareness . . . is at the core": Dolan 2006, p. 85.

129 "So this neuron fired": Gallese, Columbia symposium talk.

129 Charles Darwin spent hundreds of pages: Darwin 1955. For example, he described the musculature and expressions associated with blushing, astonishment, surprise, horror, fear, laughter, contempt, "sneering and defiance," and even meditation, of which he wrote, "here we have another instance of movement round the eyes in relation to the state of mind" (p. 227).

129 Imitation and mimicry: Richerson and Boyd 2005.

130 "Perhaps in our lineage": Ibid., p. 138.

130 She has been conducting experiments: For recent research on the neuroscience of empathy, see Singer 2006; Vignemont and Singer 2006.

130 "seems to suggest that if the system": Gallese, Columbia symposium talk.

131 Not necessarily an innate *abundance:* Gallese makes the point that mirror neurons are "a necessary but not sufficient condition for compassion" (personal communication, August 27, 2009) and adds that imitation can lead to both empathy and violence (see Gallese 2009).

131 Davidson has recruited laypeople to undertake: Lutz et al. 2008.

131 neural correlates of admiration and compassion: Immordino-Yang et al. 2009.

131 "even just a few minutes": Hutcherson, Seppala, and Gross 2008.

131 "Are there strategies that may influence": Davidson, NYU colloquium talk.

132 "It seems odd": Matthieu Ricard, personal interview.

CHAPTER EIGHT: HUMILITY

133 "They were the simple belongings": Sulzberger and Chan 2009.

133 In the fall of 1888: Fischer 1950, pp. 23–24; Gandhi 1960, pp. 69–70.

133 *gandhi* means "grocer": Fischer 2002, p. 4.

134 "lest anyone should poke fun": Ibid., p. 6.

134 "the all too impossible task": Ibid., p. 21.

134 "was wearing at the time": Fischer 1950, p. 24.

134 "become polished": Fischer 2002, p. 28.

134 "If my character made a gentleman out of me": Ibid.

135 "My hesitancy in speech": Ibid., p. 24. For someone hesitant of speech, it may be added, Gandhi produced an unbelievable volume of written work.

135 "annihilation of one's self": Gandhi 1986, p. 20.

136 "Truth and the like perhaps admit": Ibid., p. 146.

136 One of the earliest (and most amusing) passages: For the story of the golden tripod, see Plutarch 1960, pp. 46–47.

136 The word derives from the Latin *humilis: The Oxford English Dictionary* gives this origin. The secondary *humus* connection has been made, perhaps less reliably, on a number of Internet sites, including Wikipedia.

136 "Humility is like a vessel": Ricard 2006, p. 211.

136 "the moral agent's proper perspective": Greenberg 2005, p. 133.

137 "S is wise iff": This and other remarks come from Sharon Ryan, "Wisdom," in *The Stanford Encyclopedia of Philosophy,* available online at http://plato.stanford.edu/.

137 Moses was "very humble": Alter 2004, p. 742.

137 "compassionate, lovers of the brethren": Connolly 1963, p. 13.

137 But the dominant religious message: For the Wisdom literature in the Bible, see in particular Proverbs ("The fear of the Lord is the beginning of knowledge; fools despise wisdom and instruction," *The New Oxford Annotated Bible,* p. 905).

137 "consists in keeping oneself": *The Catholic Encyclopedia* (1910), p. 543 (the original source is Aquinas, *Summa Contra Gentiles,* book IV, ch. LV). *The Catholic Encyclopedia* is especially explicit about the religious devaluation of secular virtue; it defines humility as "lowliness or submissiveness," and it notes, "As applied to persons and things it means that which is abject, ignoble, or of poor condition, as we ordinarily say, not worth much" (p. 543). On the other hand, Norenzayan and Shariff 2008 argue that religious customs have the effect of promoting "pro-social" behaviors, including humility and altruism.

137 "Small Men": Confucius 2000, p. 123 ("A true gentleman is calm and at ease; the Small Man is fretful and ill at ease").

138 "Truth can be cultivated": Gandhi 1986, p. 146.

138 "Most people dislike vanity": Franklin 1969, p. 2.

138 There is a wonderful anecdote: Plutarch 1960, p. 169.

139 "Few people are wise enough": La Rochefoucauld quoted in Ricard 2006, p. 213.

139 They had set out to explore the relationship: Chatterjee and Hambrick 2007. For a variation on this theme, involving the square footage of the residences of CEOs, see Maremont 2007.

139 make more "extreme choices": Chatterjee and Hambrick 2007, p. 378.

140 "CEO narcissism is also related": Ibid.

140 "It all got launched by Lee Iacocca": Donald C. Hambrick, telephone interview, October 20, 2008.

140 "extreme personal humility": Collins 2001a, p. 21.

140 "possess this paradoxical combination of traits": Collins 2001b, p. 68.

140 "boards of directors frequently operate": Ibid., p. 75.

141 narcissism (but not humility): Lasch 1979, pp. 34–35. For development of the NPI, see Raskin and Hall 1979; for a popular recent account of the NPI, see Rosenbloom 2008.

141 "We know that narcissism correlates": W. Keith Campbell, telephone interview, June 18, 2009.

141 "Narcissists rate themselves highly": Chatterjee and Hambrick 2007, p. 354.

142 "Somewhat surprisingly": Exline 2008, p. 55; see also Exline and Geyer 2004.

142 June Price Tangney: Tangney 2000; Tangney 2002.

142 "A humble self-view need not be negative": Exline 2008, p. 56.

142 "Has an ocean drop an individuality": Gandhi 1986, p. 20.

142 "Humility is not self-deprecation": Templeton 1997, p. 162.

143 "Humility leads to prayers": Ibid., p. 30.

143 "To the extent that it helps people": Exline 2008, pp. 56–57.

143 "Love is the strongest force": Fischer 2007, p. 179.

143 "I adopt the change": Ibid., p. 139.

144 "So vain and frivolous a thing": Montaigne 2003, pp. 111–12.

144 "the essence of wisdom is to hold the attitude": Meacham 1990, p. 187.

144 *sagacity,* or exceptional discernment: Sternberg 1985.

145 "Interdependence is and ought to be": Fischer 2002, pp. 168–69.

145 "altered" the Constitution: Wills 1992, p. 38.

145 "undertook a new founding": Ibid., p. 39.

146 "eminently satisfactory": Abraham Lincoln to Edward Everett, November 20, 1863, in Lincoln 1989, p. 537.

CHAPTER NINE: ALTRUISM

148 As Clayton quickly realized: Vivian Clayton, personal interview, March 27, 2007.

148 But the "wisdom trajectory" of Solomon's life: The story of King Solomon involves several books in the Bible, not only 1 Kings 1–11 (which contains the main "biographical" material), but Deuteronomy (17:16–17, in which God defines the rules of kings), Proverbs (which codify some of his three thousand proverbs), and Song of Solomon (a collection of his purported love poetry). I have used the *New Oxford Annotated Bible* and the *Jewish Study Bible* for this account of Solomon's life.

148 Upon the death of King David: The succession story is in 1 Kings 1:1.

148 "If he proves to be a worthy man": 1 Kings 1:52.

149 "could not get warm": 1 Kings 1:1. On Adonijah's request and the scholarly interpretation, see the *New Oxford Annotated Bible,* p. 492n, and *Jewish Study Bible,* p. 675.

149 "very great wisdom": 1 Kings 4:29.

149 "Ask what I should give you": 1 Kings 3:5. The sequence when God bestows wisdom on Solomon during a dream is in 1 Kings 3:5–10.

150 "What is more fundamental": Ernst Fehr, personal interview, New York, January 11, 2008.

151 Charles Darwin was terrified of bees: Dugatkin 2006, pp. 1–11; Browne 1995, p. 203. Dugatkin writes (p. 5): "It is hard to overemphasize just how concerned Darwin was about the problem of sterile animals that helped others through their acts of altruism. That was simply not the way he envisioned natural selection operating, and at times, the problem of the sterile altruists would, as he himself noted, drive him 'half mad.' "

151 "special difficulty": Darwin 2003, p. 721.

151 The answer, Darwin began to grasp: For good popular versions of the biological study of altruism, see Ridley 1996; Dawkins 1989.

152 "will surely come to be seen": Curry 2006, p. 683.

152 "Is there any single saying": Confucius 2000, pp. 86–87.

152 *gen*—the central concept: For meaning of *gen,* see Lau's introduction in *Mencius* (1970), p. 12.

152 "Do to others": Luke: 6:31.

152 The version of the Greek philosopher Isocrates: Isocrates 1928, p. 111.

152 "Where gentlemen set their hearts": Confucius 2000, p. 96.

153 "Nature has implanted in the human breast": Smith 1976, p. 125. For an engaging review of Smith's prescient insights on behavioral economics, see Ashraf, Camerer, and Loewenstein 2005.

153 "Throughout most of its history": Glimcher et al. 2008, p. 215.

154 "Just like other people": Fehr 2004, p. 701. This essay, in *Nature*'s "Turning Points" series, gives a good summary of the evolution of Fehr's work on altruism and altruistic punishment.

154 "I searched for ways": Ibid.

155 In experiments that use game theory: Bowles and Gintis 2002, p. 125; Fehr and Fischbacher 2003, p. 785.

155 "strong reciprocity": Fehr and Fischbacher 2003, p. 785; Gintis 2000.

155 But burnishing one's reputation: On "signaling" reproductive desirability, see Fehr and Fischbacher 2003, p. 789.

155 In this classic game: This game-theory game and its history have been exhaustively described, including book-length treatments such as Poundstone 1992. For a compact explanation and significance, see Ridley 1996, pp. 53–57; Pfaff 2007, pp. 17–19.

156 In a study published in 2002: Rilling et al. 2002.

156 When experimental subjects play Prisoner's Dilemma: Glimcher et al. 2008; see also Judson 2008.

157 altruistic behavior in monkeys: Ashraf, Camerer, and Loewenstein 2005, p. 136.

157 In 2006, the National Football League: *Sports Business Daily,* September 6, 2007 (http://www.sportsbusinessdaily.com/article/114714).

157 The Ultimatum Game works like this: Fehr 2008, p. 218.

157 Humans have a predictable threshold: Ibid.; Knoch et al. 2006, p. 829.

158 "If there's going to be a package deal": Reeves quoted in Rooney 2007, p. 97. For the specifics of the argument and subsequent quotes in Rooney, see ibid., pp. 95–100.

158 Alan Sanfey and his colleagues: Sanfey et al. 2003.

159 "From hunter-gatherer societies to nation states": Nowak 2006, p. 1560.

159 "There is an interdependency here": Ernst Fehr, personal interview.

160 In order for a society to function cooperatively: Quervain et al. 2004, p. 1254.

160 "A key element of the enforcement": Fehr and Fischbacher 2003, p. 786.

160 "We have lived in cooperative societies": Robert Boyd, telephone interview, January 23, 2008.

161 We get a neural kick: Quervain et al. 2004, p. 1254.

161 As part of this experiment: Knoch et al. 2006; see also Fehr 2008, pp. 223–24.

162 researchers at the University of Illinois: Hsu, Anen, and Quartz 2008.

163 "unified measure of efficiency": Ibid., p. 1092.

163 "More broadly": Ibid., p. 1095.

163 In 2006, researchers in Europe: Gürerk, Irlensbasch, and Rockenbach 2006.

164 As the game unfolded: For a more detailed explanation of how this worked, see Henrich 2006.

164 "fully depopulated": Gürerk, Irlenbach, and Rockenbach 2006, p. 108.

164 "the puzzle of cooperation": Henrich 2006, p. 60.

165 When researchers run these "public goods" experiments: Ibid.

165 "decay in cooperation": Fehr and Fischbacher 2003, p. 786. For recent findings that punishers do not themselves succeed, see Dreber et al. 2008; Milinski and Rockenbach 2008.

165 "wisdom is corrective": Kekes 1983, p. 283.

165 "Wisdom is always relevant": Ernst Fehr, personal interview.

166 One of the reasons Solon won acclaim: Plutarch 1960, p. 59.

166 "The city where those who have not been wronged": Ibid., p. 60.

166 The followers of Confucius similarly delighted: Confucius 2000, p. xvi.

166 "If the emperor": Ibid.

167 "The brain is a very ingenious and flexible device": Ernst Fehr, personal interview.

167 Biblical scholars note that the king spent twice as long: On the construction of Solomon's palace, see 1 Kings 7; on God's anger and Solomon's demise, see 1 Kings 11.

167 "Since this has been your mind": 1 Kings 11:11.

167 "I will take the kingdom away": 1 Kings 11:35.

168 "Lead them; encourage them!": Confucius 2000, p. 161.

CHAPTER TEN: PATIENCE

169 "canvas-bellying breeze": Homer 1963, p. 214.

169 "will sing his mind away": Ibid., p. 210.

170 "unexpected warp": Ainslie 2001, p. 33.

171 Jon Elster, a psychologist: Elster 1979. On page 36 of *Ulysses and the Sirens,* Elster brilliantly points out that by resorting to the binding, "Ulysses was not fully rational, for a rational creature would not have to resort to this device. . . . His predicament—being weak and knowing it—points to the need for a theory of imperfect rationality that has been all but neglected by philosophers and social scientists." The notion of being weak and knowing it (self-awareness), combined with a behavioral strategy that values future outcomes more than immediate temptations (self-control), provides a nice shorthand version of wisdom. See also Elster 2009.

171 "sets a person against herself": Ainslie 2001, p. 62.

172 referred to him as a "genius": David Laibson, telephone interview, February 14, 2009.

172 "one of the most interesting and creative people": Frank 2007, p. vii.

172 Leading neuroscientists with whom I spoke: Nathaniel Daw, personal interview, New York, April 16, 2008; Peter Dayan, telephone interview, August 3, 2008.

173   "Self-control . . . is really the art": George Ainslie, personal interview, Coatesville, Pennsylvania, March 11, 2009.

173   "how people knowingly choose": Ainslie 2001, p. 3.

174   A series of unexpected results: George Ainslie, telephone interview, August 18, 2008; Ainslie 2001, p. x.

174   Ainslie first created a simple maze: George Ainslie, telephone interview, May 2, 2008.

175   "the matching law": Rachlin and Laibson 1997.

175   "I went to Dick": George Ainslie, personal interview.

176   The reason this graph of our motivation: An example of a hyperbolic discount curve can be seen below in the illustration here (based on Elster 1999, p. 171). What makes it so powerful, in terms of human behavior, is the way the perceived value of a smaller, short-term reward exceeds the value of a larger, long-term reward as a person (or pigeon, or rat) gets closer to the short-term reward. The basic idea is that the value of Option B (delayed, bigger reward) is larger than A at time point T_1, but *smaller* at time point T_2 because of the nature of the curves.

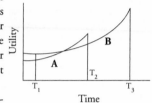

176   In a large number of experiments: Reviewed in Glimcher et al. 2008.

177   "Do not the same magnitudes": Socrates quoted in Ainslie 2001, p. 4.

177   "standing so close to a shorter building": George Ainslie, telephone interview, August 18, 2008.

177   by the time Ainslie published the results: Ainslie 1975.

177   "The irony of smart people doing stupid things": Ainslie 2001, p. 27.

177   "It usually takes some kind of effort": Ibid., p. 36.

177   "I would say the main purpose of wisdom": George Ainslie, telephone interview, August 18, 2008.

178   "seduction by short-term rewards": Ainslie 2001, p. 65.

178   "mighty waves of temptation": Saint Augustine 2001, p. 17.

178   "became to myself a land of famine": Ibid., p. 42.

178   "earthly things": Ibid., p. 49.

178   "The love of wisdom": Ibid.

179   "his words aroused me": Ibid.

179   "I *want* to have a love affair": Ainslie 2001, p. 14.

179   "the expected value of your future self-control": Ibid., p. 289.

179   "patience is companion": Augustine 1847, p. 546.

180   "is and ought only to be": Hume 1978, p. 415.

180   Those experiments began to become possible: George Ainslie, telephone interview, August 18, 2008.

181   "quasi-hyperbolic time-discounting function": Laibson 1997, p. 443.

181   "It makes a lot of these concepts usable": David Laibson, telephone interview. For a lay account of Laibson's mathematical work, see Cassidy 2006.

181   "Most of the benefits of exercise": David Laibson, telephone interview.

181   "George was the first psychologist": Ibid.

182   Their working hypothesis: McClure et al. 2004, p. 504.

182   "In a field one summer's day": I've used Jacob's 1894 translation of "The Ant and the Grasshopper" (http://en.wikisource.org/wiki/The_Ants_and_the_Grasshopper), but the popular de La Fontaine version (de La Fontaine 2004, p. 9) is much more aptly economic in its language. The grasshopper tells the ant, "I'll repay you, never fear, / Honest insect,

ere the fall, / Interest and principal." But it continues, "One fault from which the Ant is free / Is making loans too readily."

183   "Human decision makers seem to be torn": McClure et al. 2004, p. 504.

184   "the idiosyncrasies of human preferences": Ibid., p. 506.

184   a group from New York University: For the NYU study, see Kable and Glimcher 2007.

184   younger people are steeper discounters: On age and related temperamental differences in temporal discounting, see Ainslie 2001, p. 34.

185   In 2007, the Princeton group published the results: McClure et al. 2007; see also Lamy 2007.

185   "way too simplistic": Stephen Kosslyn, telephone interview, August 4, 2008.

185   their results "falsify": Kable and Glimcher 2007, p. 1631, where they state several times that their results "falsify" the hypothesis of McClure et al. 2004. See also Glimcher, Kable, and Louie 2007 and, for the countervailing argument on multiple systems, Sanfey and Chang 2008.

185   "There's no complex behavior": Stephen Kosslyn, telephone interview.

185   "This work is an important step": Ainslie and Monterosso 2004, p. 421.

186   But doctors have recently observed: Michael J. Frank, telephone interview, April 15, 2008.

186   "predicts that when faced": Frank et al. 2007, p. 1309.

186   "I think of it as a basic mechanism": Michael J. Frank, telephone interview.

187   the evolution of patience: See Stevens, Hallinan, and Hauser 2005; Rosati et al. 2007.

187   "less willing to wait for food rewards": Rosati et al. 2007, p. 1663.

187   "humans are more willing to wait": Ibid.

188   "The only reason we're not pure calculating machines": George Ainslie, telephone interview, August 18, 2008.

188   "Wisdom as it applies to self-control": Ibid.

CHAPTER ELEVEN: DEALING WITH UNCERTAINTY

190   "I bet I could beat him": This and other remarks by Cohen, unless otherwise noted, are from Jonathan D. Cohen, personal interview, Princeton, New Jersey, October 31, 2007.

192   "In this world second thoughts": Euripedes, *Hippolytus,* quoted in Fitts 1947, p. 261.

192   "A Grandmaster . . . makes the best moves": Kasparov 2007, p. 26.

193   "fast and deep calculation": Ibid., p. 45.

193   "a well-defined form of computation": Poundstone 1992, p. 6.

193   "All is flux": This is a common variation attributed to Heraclitus, as in Plato's *Cratylus* (402A).

194   "Most in the audience were aware": Matson 2006, p. 15. As Matson points out in "Certainty Made Simple," humans are perhaps the only animals that base actions on beliefs as well as trial-and-error experience ("belief" in the sense that we believe it when someone else tells us, in person or in a book, facts that shape our behavior); he makes the more interesting point that even false beliefs (including imagination about the future) can motivate thought and behavior, a philosophical echo of George Ainslie's notion that imagination is crucial to willpower.

194   "The greatest scholars are not the wisest men": Montaigne 2003, p. 118.

194   Psychological scholars of wisdom: Baltes 2004, p. 28; Birren and Svensson, 2005, p. 24; Ardelt 2005, b, p. 7.

194   "Twice is quite enough": Confucius, 2000, p. 104.

195   "if one thinks more than twice": Ibid.

195   "Repeatedly, dwell on the swiftness": Marcus Aurelius 1992, p. 11. Marcus Aurelius strikes much the same theme in the beautiful conclusion to the second meditation: "Of man's

life, his time is a point, his existence a flux, his sensation clouded, his body's entire composition corruptible, his vital spirit an eddy of breath, his fortune hard to predict, his fame uncertain. Briefly, all the things of the body, a river; all the things of the spirit, dream and delirium; his life a warfare and a sojourn in a strange land, his after-fame oblivion. What then can be his escort through life? One thing, and one thing only, Philosophy." In streetwise words, a love of wisdom.

195 "a potentiation of common sense": Medawar 1984, p. 110.

196 "When one hears a maxim": Confucius 2000, pp. 148–49.

198 On October 27, 1962: Dobbs 2008, pp. 190–94, 203–5.

198 "persuaded him to calm down": Ibid., p. 317.

198 "introspective, skeptical nature": Ibid., p. 225.

198 "forever questioning conventional wisdom": Ibid.

198 "The evolution of our emotional apparatus": Cohen 2005, p. 12. The basic theory was laid out in Miller and Cohen 2001.

199 "a broad range of decisions": Cohen 2005, p. 3.

199 "countervailing habits or reflexes": Ibid., p. 10.

200 "Vulcanization . . . is the process of treating a substance": Ibid., p. 19.

200 "protect us against ourselves": Ibid.

201 "Steep discounting may have been highly adaptive": Ibid., p. 18.

201 Ainslie attributes a good part: George Ainslie, telephone interview, August 18, 2008.

201 "a potentially fundamental instability": Cohen 2005, p. 21.

202 Many neuroscientists remain skeptical: Part of the "mulitiple system" debate is sociological and involves different experimental approaches to human decision making. In a footnote to his "Vulcanization" paper, Cohen provides a laundry list of objections to certain experimental approaches in neuroeconomics, clearly intended to subvert classical and widely accepted economic notions of optimal behavior. Among the points he raises: Can maximizing the optimal behavior of the individual in some way reduce or sabotage the optimal behavior of the group? What's the temporal component of optimal decision making: Should we opt for benefit immediately, tomorrow, next year, or over a lifetime? And can a decision be optimal while at the same time not good for the decider in the strictest economic sense? In other words, once we stray off the reservation in terms of an experimentally strict, economics-based definition of optimal behavior, the concept of a "good decision" suddenly gets very complicated. Cohen acknowledges this when he speaks of how some decisions "deviate from optimality."

202 Peter Dayan: Dayan 2008; Peter Dayan, telephone interview, August 3, 2008.

202 the downside, recent research is beginning to show: Dayan 2008.

202 "He that is really wise": Confucius 2000, p. 176; on Socrates and perplexity, see Jaspers 1962, p. 8. (In his dialogues with Meno and Theaetetus, Socrates admits that he "infects" other people with perplexity, and tells Theaetetus that "dizziness" is the beginning of philosophy—of wisdom, in short.)

203 Confucius made the point: "To go too far . . ." in Confucius 2000, p. 147.

203 "the experience that knew": Marcus Aurelius 1992, p. 3.

203 In strictly scientific terms: Cohen, McClure, and Yu 2007.

203 "The need to balance exploitation": Ibid., p. 933.

204 Cohen and his colleagues review some recent evidence: Ibid., p. 937.

205 "People do have the capacity": Cohen 2005, p. 3.

205 the only scientist I heard utter the word *deliberation:* Antonio Damasio, "Emotions, Decision Making and Social Cognitions," Symposium on "Neuroeconomics: Decision-

Making and the Brain," New York University, January 11, 2008. In his remarks at this meeting, Damasio seemed to subscribe to a multiple system of valuation, if not decision making. "We humans, for better or worse, have two systems of value." One, he said, is clearly nonconscious and based in emotion; the other operates in the cerebral cortex. But he also emphasized that these are not separate, but, rather, intertwined systems when he said, "Emotion is in the loop of reason." "The beauty, and the great problem, is that both systems in human beings are in operation at all times. And at one time or another, one will prevail upon the other. And it's not a matter that one is better than the other, but that we have to live with both."

206 "There is nothing certain but uncertainty": Montaigne 2003, p. 563.

206 "I've spent more time at Bear Stearns": Lattman and Strasburg 2008.

207 "Happy are those who find wisdom": Proverbs 3:13–14.

207 "proud ambition": Smith 1976, p. 62.

CHAPTER TWELVE: YOUTH, ADVERSITY, AND RESILIENCE

211 "exemplifies the antithesis of a nurturing environment": These and other autobiographical remarks come from Mario R. Capecchi, "The Making of a Scientist," acceptance speech for the 1996 Kyoto Prize (a slightly edited version, from which these quotes were taken, appears on the Howard Hughes Medical Institute Web site: http://www .hhmi.org/news/nobel20071008a.html. A similar but updated account is Capecchi 2008; it appears on the Web page of the Nobel Foundation: http://nobelprize.org/cgi-bin/ print?from=%2Fnobel_prizes%2Fmedicine%2Flaureates%2. Additional details and quotes appear in citations hereafter and in Stix 2008; Gumbel 2007; Collins 2007; Sample 2007.

212 Two years after Mario's birth: Capecchi learned only in 2007, through news reports, about the existence of his half sister; see http://www.msnbc.com/id/25007775.

212 scene of breathtaking poverty: Morante 1977.

213 "she lived essentially in a world of imagination": See Sample 2007.

214 Aristotle had a speech impediment: On Aristotle, see Matson 1987a, pp. 111, 114; on Moses, see Exodus 4:10–17; on Socrates, see Jaspers 1962, p. 5; on Pericles' "deformity," see Plutarch 1960, p. 167; on Gandhi's childhood, see Fischer 2002, p. 11; on Confucius, see Jaspers 1962, p. 41; on Lincoln's mother, see Shenk 2005, pp. 13–14; on Buddha, see Strong 2001, p. 44.

215 "Curiously, of the dozen or so figures": Strathern 1999, p. 13.

215 A number of psychologists who have dared to tackle: Baltes and Staudinger 2000; Ardelt 2005b.

215 In no fewer than four different studies: These are summarized in Baltes and Staudinger 2000, p. 128; Staudinger 1999. See also Pasupathi, Staudinger, and Baltes 2001.

215 "has suggested that the major period": Baltes and Staudinger 2000, p. 128.

216 Second, wisdom often grew: Jacqui Smith, Institute for Social Research, University of Michigan, Ann Arbor, telephone interviews, March 23, 2007, and March 27, 2009.

216 "In the Berlin Aging Study": Jacqui Smith, telephone interview, 2007.

216 In his valedictory work on wisdom: Baltes 2004, pp. 179–83.

216 "there is good reason to assume": Ibid., p. 182.

217 Many elderly people who scored high: Ardelt 2005b.

217 "Adversity early in life": Monika Ardelt, telephone interview, March 7, 2007.

217 "It is a rough road": Seneca 1962, p. 285.

218 They arranged for twenty pairs of the primates: Parker et al. 2004; Parker et al. 2006.

218 "stress inoculation": Parker et al. 2004, p. 933.

218 "intermittent stress inoculation": Ibid., p. 934. "Intermittent" is an important part of the concept; Parker believes that, like booster shots in traditional immunizations, repeated exposure to stress at regular intervals is crucial to the effect.

219 By the end of this elaborate yearlong experiment: Ibid., pp. 938–40.

219 "These are all indications of dimished anxiety": Karen Parker, telephone interview, March 5, 2009.

219 "These findings suggest that prior experience": Parker et al. 2004, p. 938.

220 Moreover, follow-up research: Parker et al. 2005.

220 "Our stress-inoculated monkeys": Karen Parker, telephone interview.

220 Unpublished data suggest: Ibid. An earlier experiment by the same group (Lyons et al. 2002) showed greater prefrontal development on the basis of early experience.

220 Bruce McEwen: Tang et al. 2003.

221 Researchers at the State University of New York Downstate: Kaufman et al. 2007.

221 Beginning around 1970: Cited in Parker et al. 2004, p. 934.

221 Nearly thirty years ago, Richard Davidson: Davidson and Harrington 2002.

221 In a fascinating study published back in 1989: Davidson and Fox 1989.

222 "little evidence of stability": Davidson and Harrington 2002, p. 122.

222 "This is a period": Ibid.

222 "One of the major challenges": Ibid., p. 121.

222 A depressingly deep psychological literature: For a review, see Curtis and Cicchetti 2003.

222 A beautiful set of animal experiments: Weaver et al. 2004.

223 "type, timing, duration, and severity": Parker et al. 2004, p. 939.

223 As Laura Carstensen's elegant longitudinal studies: Carstensen et al. 2000.

223 And as Fredda Blanchard-Fields's work: Blanchard-Fields 2007.

223 it seems plausible that resilience: It is also possible, as Paul Glimcher points out, that greater emotional evenness later in life could be a fortuitous side effect of age-related neurological decline (Paul Glimcher, personal interview, New York, August 28, 2008).

223 Whether that mastery is partly acquired: Lomranz 1998.

224 "Time is a game played beautifully": Heraclitus 2001, p. 51.

224 "the lessons of noble character": Marcus Aurelius 1992, p. 1.

225 "the condition of the human race": Gibbon 1932, p. 70.

225 as historians have been forced to concede: On Marcus Aurelius's unwise judgment, see ibid., pp. 73–74; Marcus Aurelius 1992, pp. vii–xi.

225 "The monstrous vices of the son": Gibbon 1932, p. 74.

225 "several inconsistencies": Barry and Foy 2007.

225 "three times, each for a duration": In the updated Nobel account, Capecchi also revealed that his father physically abused him.

226 "Capecchi took an almost scholarly interest": Barry and Foy, 2007.

CHAPTER THIRTEEN: OLDER *AND* WISER

227 Let us say you are a woman of a certain age: This anecdote is based on Blanchard-Fields 2007.

228 What she has found: For recent findings, see Blanchard-Fields, Mienaltowski, and Seay 2007; Blanchard-Fields, Stein, and Watson 2004; Watson and Blanchard-Fields 1998.

228 Adults between the ages of sixty and eighty: Blanchard-Fields, Mienaltowski, and Seay 2007.

228 "far from diminishing": Cicero 1971, p. 220.

229 "more effective" problem solvers: Blanchard-Fields 2007, p. 26.

229 "The older woman's primary use": Ibid.

229 "the apprehension of events": Marcus Aurelius 1992, p. 12.

230 The types of decline that come with aging: For the types of cognitive decline associated with aging, see Carstensen 2007; Hedden and Gabrieli 2004; Goldberg 2005, pp. 43–48.

230 when scans of my own brain: Hall 1999b.

230 The brain loses 2 percent of its weight: Nuland 1994, pp. 55–57.

230 "The end result": Ibid., p. 57.

230 "Neuroscientists may actually have discovered": Ibid., p. 56.

230 The part of the brain most vulnerable: Goldberg 2005, p. 45.

231 "having lived longer": Baltes and Staudinger 2000, p. 128.

231 "a more broadly based decline": Ibid.

231 the Baltes group in Berlin conducted studies: Ibid. For more recent research in cognitive psychology and the role of emotion in social cognition, see Blanchard-Fields 1998, pp. 238-65.

231 Jessica Andrews-Hanna and her colleagues: Andrews-Hanna et al. 2007.

231 This combination of reasonably intact hardware: Mahncke et al. 2006.

232 The neuroscientist Elkhonon Goldberg has proposed: Goldberg 2005, p. 19.

233 In the late 1930s, physicians at Harvard University: The early history of the Grant Study is described in Vaillant 1977, pp. 3–12. For a recent popular account of the study, see Shenk 2009.

234 "There's a lot of wisdom in there": Vivian Clayton, personal interview, Orinda, California, March 27, 2007. As Vaillant wrote in *Adaptation to Life* (p. 11), "Their lives were too human for science, too beautiful for numbers, too sad for diagnosis and too immortal for bound journals."

234 "people who are well": Vaillant 1977, p. 3.

234 "ego mechanisms of defense": Ibid., pp. 75–77.

234 "a man's adaptive devices": Ibid., p. 19.

234 "an alteration of the user's nervous system": Ibid., p. 81.

235 in what he called "defense mechanisms": I am indebted to George Ainslie (personal interview, March 11, 2009) for elucidating these issues; Ainslie credits Freud as being "the first author to conceive internal motivational conflict in economic terms" (Ainslie 1989, p. 11). He also makes the point that Jane Loevinger deserves credit for creating the hierarchy of defense strategies that forms the scaffold of interpretation in *Adaptation to Life*.

235 "denial, distortion, and projection": Vaillant 1977, p. 78.

235 "altruism, humor, suppression, anticipation, and sublimation": Ibid., p. 84.

236 "channeled rather than dammed or diverted": Ibid., p. 386.

236 "these mechanisms integrate": Ibid., p. 85.

236 "were far better equipped": Ibid., p. 86.

236 "David Goodhart": Goodhart is described ibid., pp. 15–29. Subsequent quotations about Goodhart are taken from this discussion.

237 "isolated traumatic events": Ibid., p. 368.

237 "characteristic reaction patterns": Ibid., p. 369.

237 "It is effective adaptation to stress": Ibid., p. 374. For a more academic version of some of the same material, see Vaillant 1993. On page 7 of *The Wisdom of the Ego*, Vaillant writes, "The mind's defenses—like the body's immune mechanisms—protect us by providing a variety of illusions to filter pain and to allow self-soothing." He goes on to note, "The wisdom of the ego referred to in the title of this book is not the wisdom of vanity but the wisdom of the integrated adaptive central nervous system."

237 "wisdom involves the toleration": Vaillant 2003, p. 256.

237 their analysis of the Harvard men: Vaillant et al. 2008.

237 In a second, ongoing study: Monika Ardelt, telephone interview, March 7, 2007.

238 In Blanchard-Fields's useful idiom: For "to do" and "let it be," see Blanchard-Fields 2007, p. 28; on spouse having an affair, see Blanchard-Fields 1998, p. 256.

239 If, as neuroscientists like Elkhonon Goldberg: For his argument on wisdom and pattern recognition, see Goldberg 2005, pp. 149–60.

239 An experimental bias: Lisbeth Nielsen, National Institute on Aging, telephone interview, September 5, 2008.

240 "What is real wisdom?": This and subsequent Erikson quotes in Goleman 1988.

CHAPTER FOURTEEN:
CLASSROOM, BOARDROOM, BEDROOM, BACK ROOM

241 Deep into his singular adventure in parenting: Isaacson 2003, pp. 10–35; Van Doren 1938, pp. 5–13; Franklin 1969, pp. 1–8.

242 "His great excellence": Franklin 1969, p. 8.

242 "he was also much consulted": Ibid.

242 "some ingenious or useful topic": Ibid.

242 "perfect inattention": Ibid.

242 "Be neither silly, nor cunning": Franklin 1997, p. 1192.

242 "unknown to fame": Ibid., p. 16.

243 "What is more interesting": This and other 2007–2008 Tufts wisdom essay questions available at http://admissions.tufts.edu/print.php?pid=186.

243 "I find them to be remarkably effective pieces": This and subsequent Coffin quote from Lee Coffin, telephone interview, November 7, 2008.

244 "All the advantage you recognize": Montaigne 2003, p. 123.

244 The definition, in a nutshell: Sternberg 1990, pp. 142–59.

245 "What differentiates really great leaders": This and other remarks, unless otherwise noted, are from Robert J. Sternberg, telephone interview, March 15, 2007.

245 "What matters is not *only*": Sternberg, Reznitskaya, and Jarvin 2007, p. 144.

245 editing the first formal academic text: *Wisdom: Its Nature, Origins, and Development* (1990).

245 "the application of intelligence": Sternberg, Reznitskaya, and Jarvin 2007, p. 145.

246 "a complex model accounting for real behaviors": Ibid., p. 150.

246 In the Tufts system: Sternberg, Reznitskaya, and Jarvin 2007, pp. 148–50; Linda Jarvin, telephone interview, February 28, 2008.

247 "will take a much more Socratic approach": Sternberg, Reznitskaya, and Jarvin 2007, p. 149.

247 "fabulous": This and subsequent remarks are from Marilyn Hamot Ryan, telephone interview, June 26, 2008.

248 "Diligence is the Mother": Franklin 1997, p. 1200.

248 "systematic use of information": Davenport quoted in Lawton 2007, p. B2.

249 "It's how committed are a company's senior executives": Ibid.

249 "he finds meaning in numbers": Nocera 2008, p. 185.

249 "truly great investing requires": Ibid., p. 184.

249 "is never, ever ruffled": Ibid., p. 185.

250 "The secret of managing": Thorn and Palmer 1989, p. 539.

250 "Get as much as possible done": Confucius 2000, p. 161.

250 "Promote those you know": Ibid.

251 "A gentleman takes as much trouble": Ibid. p. 97.

251 Researchers at New York University: Greenberg 2007.

251 "doing good is apparently good": Ibid.

251 "A philosophy from top management": Ibid.

251 "John D. Rockefeller spent every waking hour": Michael Castine, telephone interview, October 31, 2008.

252 "personality" of the three laboratories: For "happy" and "unhappy" labs, and an account of the insulin race, see Hall 1987.

252 Genentech continues to rate near the top: In its annual survey of top employers in the field, the business office of *Science* magazine has ranked Genentech in first place in six of the last seven years (Gwynne 2008).

252 one of its lab chiefs was hauled before Congress: Hall 1987, pp. 169–73.

253 some of the conflicts have produced litigation: Barinaga 1999; Egelko and Tansey 2008.

254 Vivian Clayton first became fascinated: Vivian Clayton, personal interview, Orinda, California, March 27, 2007.

255 "the man who plays alone": Dolci 1970, p. 4.

257 "The best way to win votes": Brooks 2007.

257 "cold, rational decisions": Brooks 2008.

257 "leave of absence": Plutarch 1960, p. 47.

257 "The attainment of justice and happiness": Revel and Ricard 1999, p. 216.

258 "An Inquiry into the Persistence of Unwisdom": Tuchman 1979.

258 "wise policy can only be made": Ibid., p. 7.

258 "Government remains the paramount field": Ibid., p. 8.

259 "fitness of character": Ibid., p. 9.

259 As *The Wall Street Journal* reported: Alter 2007.

259 "a pronounced shift in activity": Ibid.

259 But if you believe George Lakoff: Lakoff 2008.

259 backed by decades of neuroscientific and psychological experimentation: Kahnemann 2002.

260 "It's not just the word 'elephant'": Lakoff 2008, pp. 223–34.

260 the Democratic Party rushed to adopt: Bai 2005.

261 many fewer know that the experiments: Iacoboni 2008, pp. 239–43.

261 neurological picture of a political brain: Amodio et al. 2007.

261 "are more responsive to informational complexity": Ibid., p. 1246.

261 "elucidate how abstract, seemingly ineffable constructs": Ibid., p. 1247.

262 "Brain stem gone": The clinical signs of dementia exhibited by Ronald Reagan are described in Goldberg 2005, pp. 61–63.

262 "too much class hatred and bloodshed": Tuchman 1979, p. 5.

263 "When the Way prevails under Heaven": Confucius 2000, p. 127.

263 "the wise man should withdraw": Montaigne 2003, pp. 103–4.

263 "Isn't it better to be doing this": Heraclitus quoted ibid., p. 120.

CHAPTER FIFTEEN: DARE TO BE WISE

265 in Renaissance Italy: Rice 1958, pp. 48–49.

266 Socrates "ironically explains": Jaspers 1962, p. 7.

267 "recovering jerk": Pausch 2008, p. 116.

268 "Cicero says that to philosophize": Montaigne 2003, p. 67.

268 "is where our modern thinking about death": Barnes 2008, p. 40.

268 "without doubt the most noteworthy action": Montaigne 2003, p. 556.

268 "It is no great thing": Ibid., p. 561.

270 "A map of the world": Wilde 1931, p. 28.

271 "secret currents of might and mind": Emerson 1979, p. 195.

271 John Meacham found wisdom: Meacham 1990, pp. 181–83.

271 Stanley Falkow: Falkow 2008.

271 Edward O. Wilson has spent a career: Wilson 1994.

271 You'd be surprised how often bees: See Aristotle 1984, p. 975; Marcus Aurelius 1992, p. 43; Emerson 1979, p. 135; Darwin 2003, p. 721. According to Dugatkin 2006, Darwin's children helped him perform experiments with bees.

272 "Oddly enough . . . in insect life": Vivian Clayton, personal interview, Brooklyn, New York, November 20, 2007. In her drosh, Clayton wrote, "The Queen does not, by nature, reveal herself to you . . . she remains hidden, elusive. And yet, the best way to find her is not to 'look' directly. . . ."

BIBLIOGRAPHY

Abbott, A. 2009. "Brain Imaging Studies Under Fire." *Nature* 457:245.

Ainslie, George. 1975. "Specious Reward: A Behavioral Theory of Impulsiveness and Impulse Control." *Psychological Bulletin* 82:463–96.

———. 1989. "Freud and Picoeconomics." *Behaviorism* 17:11–18.

———. 1992. *Picoeconomics: The Strategic Interaction of Successive Motivational States Within the Person.* Cambridge, U.K.: Cambridge University Press.

———. 2001. *Breakdown of Will.* Cambridge, U.K.: Cambridge University Press.

Ainslie, George, and John Monterosso. 2004. "A Marketplace in the Brain?" *Science* 306:421–23.

Alter, A. 2007. "Reading the Mind of the Body Politic." *Wall Street Journal,* December 14.

Alter, A. L., D. M. Oppenheimer, N. Epley, and R. N. Eyre. 2007. "Overcoming Intuition: Metacognitive Difficulty Activates Analytic Reasoning." *Journal of Experimental Psychology: General* 136:569–76.

Alter, Robert. 2004. *The Five Books of Moses: A Translation with Commentary.* New York: Norton.

Amodio, D. M., J. T. Jost, S. L. Master, and C. M. Yee. 2007. "Neurocognitive Correlates of Liberalism and Conservatism." *Nature Neuroscience* 10:1246–47.

Andrews-Hanna, J. R., A. Z. Snyder, J. L. Vincent, C. Lustig, D. Head, M. E. Raichle, and R. L. Buckner. 2007. "Disruption of Large-Scale Brain Systems in Advanced Aging." *Neuron* 56:924–35.

Appiah, Kwame Anthony. 2008. *Experiments in Ethics.* Cambridge, Mass.: Harvard University Press.

Ardelt, M. 2000. "Antecedents and Effects of Wisdom in Old Age." *Research on Aging* 22:360–94.

———. 2004. "Where Can Wisdom Be Found? A Reply to the Commentaries by Baltes and Kunzmann, Sternberg, and Achenbaum." *Human Development* 47:304–7.

———. 2005a. Foreword to Sternberg and Jordan 2005, pp. xi–xvii.

———. 2005b. "How Wise People Cope with Crises and Obstacles in Life." *ReVision* 28:7–19.

Ariely, Dan. 2008. *Predictably Irrational: The Hidden Forces That Shape Our Decisions.* New York: HarperCollins.

Aristotle. 1908. *Nicomachean Ethics.* Trans. W. D. Ross. Oxford: Clarendon Press.

———. 1984. *The Complete Works of Aristotle (The Revised Oxford Translation).* 2 vols. Ed. Jonathan Barnes. Princeton, N.J.: Princeton University Press.

———. 2004. *The Nicomachean Ethics.* Rev. ed. Trans. J. A. K. Thomson. New York: Penguin.

Armstrong, Karen. 2001. *Buddha.* New York: Penguin.

Ashraf, N., C. F. Camerer, and G. Loewenstein. 2005. "Adam Smith, Behavioral Economist." *Journal of Economic Perspectives* 19:131–45.

Associated Press. 2008. "Nobel Prize Helps Scientist Find Lost Sister." June 6.

Augustine. 1847. *Seventeen Treatises of S. Augustine, Bishop of Hippo.* Oxford: J. H. Parker.

———. 2001. *The Confessions.* Trans. Philip Burton. New York: Everyman's Library/Knopf.

Ayer, A. J., and Jane O'Grady, eds. 1992. *A Dictionary of Philosophical Quotations.* Malden, Mass.: Blackwell.

Bai, Matt. 2005. "The Framing Wars." *New York Times Magazine,* July 17.

Baltes, Paul B. 1999. "Autobiographical Reflections: From Developmental Methodology and Lifespan Psychology to Gerontology." In *A History of Geropsychology in Autobiography,* ed. J. E. Birren and J. J. F. Schroots, pp. 7–26. Washington, D.C.: American Psychology Association.

———. 2004. "Wisdom as Orchestration of Mind and Virtue," unpublished ms. posted on the Web site of the Max Planck Institute for Human Development, Berlin, http://library.mpib-berlin.mpg.de/ft/pb/PB_Wisdom_2004.pdf (accessed January 2, 2009).

Baltes, Paul B., and U. Kunzmann. 2004. "The Two Faces of Wisdom: Wisdom as a General Theory of Knowledge and Judgment About Excellence in Mind and Virtue vs. Wisdom as Everyday Realization in People and Products." *Human Development* 47:290–99.

Baltes, Paul B., and J. Smith. 2008. "The Fascination of Wisdom: Its Nature, Ontogeny, and Function." *Perspectives on Psychological Science* 3:56–64.

Baltes, Paul B., and U. M. Staudinger. 2000. "Wisdom: A Metaheuristic (Pragmatic) to Orchestrate Mind and Virtue Toward Excellence." *American Psychologist* 55:122–36.

Barinaga, M. 1999. "Genentech, UC Settle Suit for $200 Million." *Science* 286:1655.

Barnes, Julian. 2008. *Nothing to Be Frightened Of.* London: Jonathan Cape.

Barry, C., and P. Foy. 2007. "Story of Nobelist's Past Is Inconsistent with Data." *Washington Post,* November 7, p. A16.

Begley, Sharon. 2007. *Train Your Mind, Change Your Brain: How a New Science Reveals Our Extraordinary Potential to Transform Ourselves.* New York: Ballantine.

Bateson, M., D. Nettle, and G. Roberts. 2006. "Cues of Being Watched Enhance Cooperation in a Real-World Setting." *Biology Letters* 2:412–14.

Birren, J. E., and C. M. Svensson. 2005. "Wisdom in History." In Sternberg and Jordan 2005, pp. 3–31.

Blanchard-Fields, F. 1998. "The Role of Emotion in Social Cognition Across the Adult Life Span." In *Annual Review of Gerontology and Geriatrics,* Vol. 17: *Focus on Emotion and Adult Development,* ed. K. W. Schaie and M. P. Lawton, pp. 238-65. New York: Springer.

———. 2007. "Everyday Problem Solving and Emotion: An Adult Developmental Perspective." *Current Directions in Psychological Science* 16:26–31.

Blanchard-Fields, F., A. Mienaltowski, and R. B. Seay. 2007. "Age Differences in Everyday Problem-Solving Effectiveness: Older Adults Select More Effective Strategies for Interpersonal Problems." *Journal of Gerontology* 62B:P61–P64.

Blanchard-Fields, F., R. Stein, and T. L. Watson. 2004. "Age Differences in Emotion-Regulation Strategies in Handling Everyday Problems." *Journal of Gerontology* 59B:P261–P269.

Bowles, S., and H. Gintis. 2002. *"Homo reciprocans." Nature* 415:125–27.

Brefczynski-Lewis, J. A., A. Lutz, H. S. Schaefer, D. B. Levinson, and R. J. Davidson. 2007.

"Neural Correlates of Attentional Expertise in Long-Term Meditation Practitioners." *Proceedings of the National Academy of Sciences* 104:11483–88.

Brooks, David. 2007. "Stop Making Sense." *New York Times Book Review,* August 26.

———. 2008. "How Voters Think." *New York Times,* January 18, p. A21.

Browne, E. Janet. 1995. *Charles Darwin: A Biography.* New York: Knopf.

Buchmann, Christina, and Celina Spiegel, eds. 1994. *Out of the Garden: Women Writers on the Bible.* New York: Fawcett Columbine.

Buss, K. A., J. R. Malmstadt Schumacher, I. Dolski, N. H. Kalin, H. H. Goldsmith, and R. J. Davidson. 2003. "Right Frontal Brain Activity, Cortisol, and Withdrawal Behavior in 6-Month-Old Infants." *Behavioral Neuroscience* 117:11–20.

Cacioppo, John T., Penny S. Visser, and Cynthia L. Pickett, eds. 2006. *Social Neuroscience: People Thinking About Thinking People.* Cambridge, Mass.: Bradford Books/MIT Press.

Caggiano, V., L. Fogassi, G. Rizzolatti, P. Thier, and A. Casile. 2009. "Mirror Neurons Differentially Encode the Peripersonal and Extrapersonal Space of Monkeys." *Science* 324:403–6.

Camerer, C., G. Loewenstein, and D. Prelec. 2005. "Neuroeconomics: How Neuroscience Can Inform Economics." *Journal of Economic Literature* 43:9–64.

Capecchi, Mario R. 2007. "The Making of a Scientist." Howard Hughes Medical Institute, Chevy Chase, Maryland. Available at http://www.hhmi.org/news/nobel20071008a.html.

———. 2008. "The Making of a Scientist II." Stockholm: Nobel Foundation. Available at http://capecchi.genetics.utah.edu/PDFs/163ChemBioChem.pdf.

Carroll, R. T. 2008. "The Templeton Fundies." *The Humanist,* May/June.

Cartensen, L. L. 2006. "The Influence of a Sense of Time on Human Development." *Science* 312:1913–15.

———. 2007. "Growing Old or Living Long: Take Your Pick." *Issues in Science and Technology,* Winter, pp. 41–50.

Carstensen, L. L., and C. E. Lockenhoff. 2003. "Aging, Emotion, and Evolution: The Bigger Picture." *Annals of the New York Academy of Sciences* 1000:152–79.

Carstensen, L. L., M. Pasupathi, U. Mayr, and J. R. Nesselroade. 2000. "Emotional Experience in Everyday Life Across the Adult Life Span." *Journal of Personality and Social Psychology* 79:644–55.

Carstensen, L. L., B. Turan, N. Ram, S. Schiebe, H. Ersner-Hershfield, G. Samanez-Larkin, K. Brooks, and J. Nesselroade. In press. "The Experience and Regulation of Emotion in Adulthood: Experience Sampling Across a 10-Year-Period in People Aged 18 to Over 90."

Carter, Rita. 1998. *Mapping the Mind.* Berkeley: University of California Press.

Cassidy, J. 2006. "Mind Games." *The New Yorker,* September 18.

The Catholic Encyclopedia. 1910. New York: Encyclopedia Press.

Chapman, H. A., D. A. Kim, J. M. Susskind, and A. K. Anderson. 2009. "In Bad Taste: Evidence for the Oral Origins of Moral Disgust." *Science* 323:1222–26.

Chatterjee, A., and D. C. Hambrick. 2007. "It's All About Me: Narcissistic CEOs and Their Effects on Company Strategy and Performance." *Administrative Science Quarterly* 52:351–86.

Churchland, Patricia Smith. 1986. *Neurophilosophy: Toward a Unified Science of the Mind-Brain.* Cambridge, Mass: Bradford Books/MIT Press.

Cicero. 1971. *Selected Works.* Trans. Michael Grant. New York: Penguin.

Clayton, V. 1975. "Erikson's Theory of Human Development as It Applies to the Aged: Wisdom as Contradictive Cognition." *Human Development* 18:119–28.

———. 1982. "Wisdom and Intelligence: The Nature and Function of Knowledge in the Later Years." *International Journal of Aging and Human Development* 15 (4):315–21.

Clayton, V., and J. E. Birren. 1980. "The Development of Wisdom Across the Life Span: A Reexamination of an Ancient Logic." In *Life-Span Development and Behavior, 3*, ed. P. B. Baltes and O. G. Brim, pp. 103–35. New York: Academic Press.

Clayton, V., and K. W. Schaie. 1979. "A Developmental Analysis of the Perception of Wisdom: A Multidimensional Approach." Unpublished ms.

Cohen, J. D. 2005. "The Vulcanization of the Human Brain: A Neural Perspective on Interactions Between Cognition and Emotion." *Journal of Economic Perspectives* 19:3–24.

Cohen, J. D., and K. I. Blum. 2002. "Reward and Decision." *Neuron* 36:193–98.

Cohen, J. D., S. M. McClure, and A. J. Yu. 2007. "Should I Stay or Should I Go? How the Human Brain Manages the Trade-off Between Exploitation and Exploration." *Philosophical Transactions of the Royal Society B* 362:933–42.

Collins, Jim. 2001a. *Good to Great: Why Some Companies Make the Leap . . . and Others Don't.* New York: HarperCollins.

———. 2001b. "Level 5 Leadership: The Triumph of Humility and Fierce Resolve." *Harvard Business Review*, January, pp. 66–76.

Collins, L. M. 2007. "The Quiet Man—Capecchi Is Making a Big Splash in the Genetics Pool." *Deseret Morning News*, October 8.

Confucius. 2000. *The Analects.* Trans. Arthur Waley. New York: Everyman's Library/Knopf.

Connolly, Francis X. 1963. *Wisdom of the Saints.* New York: Pocket Books.

Corcoran, K. M., and Z. V. Segal. 2008. "Metacognition in Depressive and Anxiety Disorders: Current Directions." *International Journal of Cognitive Therapy* 1:33–44.

Cotter, H. 2008. "The Glory That Was Greece from a Female Perspective." *New York Times*, December 19, p. C27.

Csikszentmihalyi, M., and K. Rathunde. 1990. "The Psychology of Wisdom: An Evolutionary Interpretation." In Sternberg 1990, pp. 25–51.

Curnow, Trevor. 1999. *Wisdom, Intuition and Ethics.* Aldershot, U.K.: Ashgate.

Curry, O. 2006. "One Good Deed." *Nature* 444:683.

Curtis, W. J., and D. Cicchetti. 2003. "Moving Research on Resilience into the 21st Century: Theoretical and Methodological Considerations in Examining the Biological Contributors to Resilience." *Development and Psychopathology* 15:773–810.

Cushman, F., L. Young, and M. Hauser. 2006. "The Role of Conscious Reasoning and Intuition in Moral Judgment." *Psychological Science* 17:1082–89.

Damasio, Antonio. 1994. *Descartes' Error: Emotion, Reason, and the Human Brain.* New York: G. P. Putnam.

———. 2003. *Looking for Spinoza: Joy, Sorrow, and the Feeling Brain.* Orlando: Harcourt.

———. 2005. "Brain Trust." *Nature* 435:571–72.

Darwin, Charles. 1955. *The Expression of the Emotions in Man and Animals.* New York: Philosophical Library.

———. 2003. *The Origin of Species and The Voyage of the Beagle.* New York: Everyman's Library/Knopf.

———. 2004. *The Descent of Man and Selection in Relation to Sex.* New York: Penguin.

Davidson, R. J., and N. A. Fox. 1989. "Frontal Brain Asymmetry Predicts Infants' Response to Maternal Separation." *Journal of Abnormal Psychology* 98:127–31.

Davidson, Richard J., and Anne Harrington, eds. 2002. *Visions of Compassion: Western Scientists and Tibetan Buddhists Examine Human Nature.* New York: Oxford University Press.

Davidson, R. J., J. Kabat-Zinn, J. Schumacher, M. Rosenkranz, D. Muller, S. F. Santorelli, F. Urbanowski, A. Harrington, K. Bonus, and J. F. Sheridan. 2003. "Alterations in Brain and Immune Function Produced by Mindfulness Meditation." *Psychosomatic Medicine* 65:564–70.

Daw, N. D., Y. Niv, and P. Dayan. 2005. "Uncertainty-Based Competition Between Prefrontal and Dorsolateral Striatal Systems for Behavioral Control. *Nature Neuroscience* 8:1704–11.

Dawkins, Richard. 1989. *The Selfish Gene.* 2nd ed. New York: Oxford University Press.

———. 2006. *The God Delusion.* Boston: Houghton Mifflin.

Dayan, P. 2008. "The Role of Value Systems in Decision Making." In *Better Than Conscious? Decision Making, the Human Mind, and Implications for Institutions,* ed. Christoph Engel and Wolf Singer, pp. 51–70. Cambridge, Mass.: MIT Press.

Dayan, P., Y. Niv, B. Seymour, and N. Daw. 2006. "The Misbehavior of Value and the Discipline of the Will." *Neural Networks* 19:1153–60.

De La Fontaine, J. 2001. *Fables.* New York: Everyman's Library/Knopf.

Delgado, M. R., M. M. Gillis, and E. A. Phelps. 2008. "Regulating the Expectation of Reward via Cognitive Strategies." *Nature Neuroscience,* published online June 29, 2008; doi: 10.1038/nn.2141.

Dijksterhuis, A., M. W. Bos, L. F. Nordgren, and R. B. van Baaren. 2006. On Making the Right Choice: The Deliberation-Without-Attention Effect." *Science,* 311:1005–7.

Dijksterhuis, A., and L. F. Nordgren. 2006. "A Theory of Unconscious Thought." *Perspectives on Psychological Science* 1 (2):95–109.

Dobbs, Michael. 2008. *One Minute to Midnight: Kennedy, Khrushchev, and Castro on the Brink of Nuclear War.* New York: Knopf.

Dolan, Ray. 2006. "The Body in the Brain." *Daedalus,* Summer, pp. 78–85.

Dolci, Danilo. 1970. *The Man Who Plays Alone.* Garden City, N.Y.: Anchor.

Dreber, A., D. G. Rand, D. Fudenberg, and M. A. Nowak. 2008. "Winners Don't Punish." *Nature* 452:348–51.

Dugatkin, Lee Alan. 2006. *The Altruism Equation: Seven Scientists Search for the Origins of Goodness.* Princeton, N.J.: Princeton University Press.

Egelko, B., and B. Tansey. 2008. "Court Reduces Genentech Damages in Royalty Suit." *San Francisco Chronicle,* April 25.

Elster, Jon. 1979. *Ulysses and the Sirens: Studies in Rationality and Irrationality.* Cambridge, U.K.: Cambridge University Press.

———. 1999. *Strong Feelings: Emotion, Addiction, and Human Behavior.* Cambridge, Mass.: MIT Press.

Emerson, Ralph Waldo. 1979. *The Essays of Ralph Waldo Emerson.* Cambridge, Mass.: Belknap Press/Harvard University Press.

Epictetus. 1998. *Encheiridion.* Trans. Wallace I. Matson, in *Classics of Philosophy,* ed. Louis P. Pojman. New York: Oxford University Press.

Erikson, Erik H. 1950. *Childhood and Society.* New York: Norton.

———. 1959. *Identity and the Life Cycle: Selected Papers.* New York: International Universities Press.

———. 1968. *Identity: Youth and Crisis.* New York: Norton.

———. 1969. *Gandhi's Truth: On the Origins of Militant Nonviolence.* New York: Norton.

Etz Hayim: Torah and Commentary. 2001. Ed. David L. Lieber. New York: The Rabbinical Assembly.

Exline, J. J. 2008. "Taming the Wild Ego: The Challenge of Humility." In *Quieting the Ego: Psychological Benefits of Transcending Egotism,* ed. J. Bauer and H. Wayment, pp. 53–62. Washington, D.C.: American Psychological Association.

Exline, J. J., W. K. Campbell, R. F. Baumeister, T. Joiner, J. Krueger, and L. V. Kachorek. 2004. "Humility and Modesty." In *The Values in Action (VIA) Classification of Strengths,* ed. C. Peterson and M. Seligman, pp. 461–75. Cincinnati: Values in Action Institute.

Exline, J. J., and A. L. Geyer. 2004. "Perceptions of Humility: A Preliminary Study." *Self and Identity* 3:95–114.

Falkow, S. 2008. "I Never Met a Microbe I Didn't Like." *Nature Medicine* 14 (10):xxvii–xxxi.

Fehr, E. 2004. "The Productivity of Failures." *Nature* 428:701.

———. 2008. "Social Preferences and the Brain." In Glimcher et al. 2008, pp. 215–32.

Fehr, E., and U. Fischbacher. 2003. "The Nature of Human Altruism." *Nature* 425:785–91.

Fischer, Louis. 1950. *The Life of Mahatma Gandhi.* New York: Harper and Bros.

———, ed. 2002. *The Essential Gandhi: An Anthology of His Writings on His Life, Work, and Ideas.* New York: Vintage.

Fitts, Dudley, ed. 1947. *Greek Plays in Modern Translation.* New York: Dial.

Flanagan, Owen. 2007. *The Really Hard Problem: Meaning in a Material World.* Cambridge, Mass.: Bradford Books/MIT Press.

Frank, M. J., J. Samanta, A. A. Moustafa, and S. J. Sherman. 2007. "Hold Your Horses: Impulsivity, Deep Brain Stimulation, and Medication in Parkinsonism." *Science* 318: 1309–12.

Frank, Robert H. 2007. *Falling Behind: How Rising Inequality Harms the Middle Class.* Berkeley: University of California Press.

Franklin, Benjamin. 1969. *Autobiography and Selected Writings.* San Francisco: Rinehart.

———. 1997. *Autobiography, Poor Richard, and Later Writings.* New York: Library of America.

Fung, H. H., and L. L. Carstensen. 2006. "Goals Change When Life's Fragility Is Primed: Lessons Learned from Older Adults, the September 11 Attacks and SARS." *Social Cognition* 24 (3):248–78.

Furuya, S., and H. Kinoshita. 2008. "Expertise-Dependent Modulation of Muscular and Nonmuscular Torques in Multi-joint Arm Movements During Piano Keystroke." *Neuroscience* 156 (2):390–402.

Galdi, S., L. Arcuri, and B. Gawronski. 2008. "Automatic Mental Associations Predict Future Choices of Undecided Decision-Makers." *Science* 321:1100–2.

Gallese, V. 2008. "Mirror Neurons and the Social Nature of Language: The Neural Exploitation Hypothesis." *Social Neuroscience* 3:317–33.

———. 2009. "The Two Sides of Mimesis: Girard's Mimetic Theory, Embodied Simulation and Social Identification." *Journal of Consciousness Studies* 16 (3):21–44.

Gallese, V., M. Rochat, G. Cossu, and C. Sinigaglia. 2009. "Motor Cognition and Its Role in the Phylogeny and Ontogeny of Intentional Understanding." *Developmental Psychology* 45:103–13.

Gandhi, Mahatma. 1960. *Gandhi's Autobiography: The Story of My Experiments with Truth.* Trans. Mahadev Desai. Washington, D.C.: Public Affairs Press.

———. 1986. *The Moral and Political Writings of Mahatma Gandhi.* Vol. II: *Truth and Non-Violence.* Ed. Raghavan Iyer. Oxford: Clarendon Press.

Gazzaniga, Michael S. 2005. *The Ethical Brain.* New York: Dana Press.

———. 2008. *Human: The Science Behind What Makes Us Unique.* New York: Ecco.

Gewirtz, P. 1996. "On 'I Know It When I See It.' " *Yale Law Journal* 105:1023–47.

Gibbon, Edward. 1932. *The Decline and Fall of the Roman Empire.* New York: Modern Library/Random House.

Gilbert, Daniel. 2005. *Stumbling on Happiness.* New York: Knopf.

Ginsberg, H. L. 1967. "Job the Patient and Job the Impatient." *Conservative Judaism* 21:12–28; reprinted in *Vetus Testamentum Supplements* 17 (1969):88–111.

Gintis, Herbert. 2000. "Strong Reciprocity and Human Sociality." *Journal of Theoretical Biology* 206:169–79.

———. 2008. "Punishment and Cooperation." *Science* 319:1345–46.

Gladwell, Malcolm. 2005. *Blink: The Power of Thinking Without Thinking.* New York: Little, Brown.

Glimcher, P. W., J. Kable, and K. Louie. 2007. "Neuroeconomic Studies of Impulsivity: Now or Just as Soon as Possible?" *American Economic Association Papers and Proceedings,* May, pp. 142–47.

Glimcher, Paul W. 2009. "Neuroeconomics and the Study of Valuation." in *The Cognitive Neurosciences,* 4th ed., ed. Michael S. Gazzaniga, pp. 1085–92. Cambridge, Mass.: MIT Press.

Glimcher, Paul W., Colin F. Camerer, Ernst Fehr, and Russell A. Poldrack, eds. 2008. *Neuroeconomics: Decision Making and the Brain.* Boston: Academic Press.

Goldberg, Elkhonon. 2001. *The Executive Brain: Frontal Lobes and the Civilized Mind.* New York: Oxford University Press.

———. 2005. *The Wisdom Paradox: How Your Mind Can Grow Stronger as Your Brain Grows Older.* New York: Gotham.

Goldin, P. R., K. McRae, W. Ramel, and J. J. Gross. 2008. "The Neural Bases of Emotion Regulation: Reappraisal and Suppression of Negative Emotion." *Biological Psychiatry* 63:577–86.

Goleman, Daniel. 1988. "Erikson, in His Own Old Age, Expands His View of Life." *New York Times,* June 14.

———. 1995. *Emotional Intelligence: Why It Can Matter More Than IQ.* New York: Bantam.

Greenberg, H. 2007. "How Values Embraced by a Company May Enhance That Company's Value." *Wall Street Journal,* October 27–28, p. B3.

Greene, Joshua D. 2003. "From Neural 'Is' to Moral 'Ought': What Are the Moral Implications of Neuroscientific Moral Psychology?" *Nature Reviews/Neuroscience* 4:847–50.

———. 2009. "The Secret Joke of Kant's Soul." In *Moral Psychology,* Vol. 3: *The Neuroscience of Morality: Emotion, Disease, and Development,* ed. W. Sinnott-Armstrong, pp. 35–80, Cambridge, Mass.: MIT Press.

Greene, Joshua D., L. E. Nystrom, A. D. Engell, J. M. Darley, and J. D. Cohen. 2004. "The Neural Bases of Cognitive Conflict and Control in Moral Judgment." *Neuron* 44:389–400.

Greene, Joshua D., R. B. Sommerville, L. E. Nystrom, J. M. Darley, and J. D. Cohen. 2001. "An fMRI Investigation of Emotional Engagement in Moral Judgment." *Science* 293:2105–8.

Grenberg, Jeanine. 2005. *Kant and the Ethics of Humility: A Story of Dependence, Corruption, and Virtue.* Cambridge, U.K.: Cambridge University Press.

Gribetz, Jessica. 1997. *Wise Words: Jewish Thoughts and Stories Through the Ages.* New York: Quill.

Gross, James J. 2008. "Emotion Regulation." In *Handbook of Emotions,* 3rd ed., ed. Michael Lewis, Jeannette M. Haviland-Jones, and Lisa Feldman Barrett, pp. 497–512. New York: Guilford Press.

Gumbel, A. 2007. "Italian Street Urchin Who Led a Global Revolution in Medicine." *The Independent,* October 9.

Gürerk, O., B. Irlenbusch, and B. Rockenbach. 2006. "The Competitive Advantage of Sanctioning Institutions." *Science* 312:108–11.

Gwynne, P. 2008. "Leadership, Stability, and Social Responsibility." *Science* 322:283–90.

Haidt, Jonathan. 2001. "The Emotional Dog and Its Rational Tail: A Social Intuitionist Approach to Moral Judgment." *Psychological Review* 108:814–34.

———. 2006. *The Happiness Hypothesis: Finding Modern Truth in Ancient Wisdom.* New York: Basic Books.

―――. 2007. "The New Synthesis in Moral Psychology." *Science* 316:998–1002.

Haidt, Jonathan, J. P. Seder, and S. Kesebir. 2008. "Hive Psychology, Happiness, and Public Policy." *Journal of Legal Studies* 37: 113S–156S.

Hall, Stephen S. 1987. *Invisible Frontiers: The Race to Synthesize a Human Gene.* New York: Atlantic Monthly Press.

―――. 1999a. "Fear Itself." *New York Times Magazine,* February 28.

―――. 1999b. "Journey to the Center of My Mind." *New York Times Magazine,* June 6.

―――. 2001. *Sound and Shadow: September, 2001.* Brooklyn, N.Y.: Privately printed; see www .stephenshall.com.

―――. 2003. "Is Buddhism Good for Your Health?" *New York Times Magazine,* September 14.

―――. 2007. "The Older-and-Wiser Hypothesis." *New York Times Magazine,* May 6.

Hauser, Marc D. 2006. *Moral Minds: The Nature of Right and Wrong.* New York: Harper-Collins.

Hauser, Marc D., F. Cushman, L. Young, R. Kang-Xing Jin, and J. Mikhail. 2007. "A Dissociation Between Moral Judgments and Justifications." *Mind & Language* 22:1–21.

Hawkes, K. 2004. "The Grandmother Effect." *Nature* 428:128–29.

Haynes, J.-D., and G. Rees. 2006. "Decoding Mental States from Brain Activity in Humans." *Nature Reviews Neuroscience* 7:523–34.

Haynes, J.-D., K. Sakai, G. Rees, S. Gilbert, C. Frith, and R. E. Passingham. 2007. "Reading Hidden Intentions in the Human Brain." *Current Biology* 17:323–28.

Hedden, T., and J. D. E. Gabrieli. 2004. "Insights into the Ageing Mind: A View from Cognitive Neuroscience." *Nature Reviews Neuroscience* 5:87–97.

―――. 2006. "The Ebb and Flow of Attention in the Human Brain." *Nature Neuroscience* 9:863–65.

Henrich, J. 2006. "Cooperation, Punishment, and the Evolution of Human Institutions." *Science* 312:60–61.

Heraclitus. 2001. *Fragments: The Collected Wisdom of Heraclitus.* Trans. Brooks Haxton. New York: Viking.

Herrmann, B., C. Thoni, and S. Gachter. 2008. "Antisocial Punishment Across Societies." *Science* 319:1362–67.

Hesiod. 1959. *The Works and Days, Theogony, and The Shield of Herakles.* Trans. Richmond Lattimore. Ann Arbor: University of Michigan Press.

Homer. 1963. *The Odyssey.* Trans. Robert Fitzgerald. Garden City, N.Y.: Anchor/Doubleday.

Horace. *The Satires and Epistles of Horace.* 1959. Trans. Smith Palmer Bovie. Chicago: University of Chicago Press.

Horgan, J. 2006. "The Templeton Foundation: A Skeptic's Take." *Chronicle of Higher Education,* April 7.

Hsu, M., C. Anen, and S. R. Quartz. 2008. "The Right and the Good: Distributive Justice and Neural Encoding of Equity and Efficiency." *Science* 320:1092–95.

Hume, David A. 1978. *A Treatise of Human Nature.* 2nd ed. Ed. L. A. Selby-Bigge. Oxford: Oxford University Press.

Hutcherson, C. A., E. M. Seppala, and J. J. Gross. 2008. "Loving-Kindness Meditation Increases Social Connectedness." *Emotion* 8:720–24.

Iacoboni, Marco. 2008. *Mirroring People: The New Science of How We Connect with Others.* New York: Farrar, Straus and Giroux.

Immordino-Yang, M. H., A. McColl, H. Damasio, and A. Damasio. 2009. "Neural Correlates of Admiration and Compassion." *Proceedings of the National Academy of Sciences* 106:8021–26.

Isaacson, Walter. 2003. *Benjamin Franklin: An American Life.* New York: Simon & Schuster.

Isocrates. 1928. *Isocrates, Vol. 1.* Trans. George Norlin. Cambridge, Mass.: Loeb Classical Library/Harvard University Press.

Iyengar, S. S., and M. R. Lepper. 2000. "When Choice Is Demotivating: Can One Desire Too Much of a Good Thing?" *Journal of Personality and Social Psychology* 79:995–1006.

James, William. 1884. "What Is an Emotion?" *Mind* 9:188–205.

———. 1981. *The Principles of Psychology.* Vols. I and II. Cambridge, Mass.: Harvard University Press.

Jaspers, Karl. 1962. *Socrates, Buddha, Confucius, Jesus: The Paradigmatic Individuals.* Trans. Ralph Manheim. San Diego: Harcourt Brace.

———. 2003. *Way to Wisdom: An Introduction to Philosophy.* 2nd ed. Trans. Ralph Manheim. New Haven, Conn.: Yale University Press/Nota Bene.

The Jewish Study Bible. 2004. Ed. Adele Berlin and Marc Zvi Brettler. New York: Oxford University Press.

Jones, R. 2007. "Learning to Pay Attention." *PloS Biology* 5:1188–89.

Joyce, James. 1976. *The Portable James Joyce.* Ed. Harry Levin. New York: Penguin.

Judson, O. 2008. "Feel the Eyes Upon You." *New York Times,* August 3.

Kabat-Zinn, Jon. 1990. *Full Catastrophe Living: Using the Wisdom of Your Body and Mind to Face Stress, Pain, and Illness.* New York: Delta.

Kable, J. W., and P. W. Glimcher. 2007. "The Neural Correlates of Subjective Value During Intertemporal Choice." *Nature Neuroscience* 10:1625–33.

Kahneman, Daniel. 2002. "Maps of Bounded Rationality: A Perspective on Intuitive Judgment and Choice." Nobel Prize lecture. Available at http://nobelprize.org.

Kant, Immanuel. 1996. *The Metaphysics of Morals.* New York: Cambridge University Press.

———. 1999. *Critique of Pure Reason.* Indianapolis: Hackett.

Kasparov, Garry. 2007. *How Life Imitates Chess: Making the Right Moves—from the Board to the Boardroom.* New York: Bloomsbury.

Kass, Leon R. 1997. "The Wisdom of Repugnance." *The New Republic,* June 2, pp. 17–26.

———. 1999. *The Hungry Soul: Eating and the Perfection of Our Nature.* Chicago: University of Chicago Press.

Kaufman, D., M. Banerji, I. Shorman, E. L. P. Smith, J. D. Coplan, L. A. Rosenblum, and J. G. Kral. 2007. "Early-Life Stress and the Development of Obesity and Insulin Resistance in Juvenile Bonnet Macaques." *Diabetes* 56:1–5.

Kekes, J. 1983. "Wisdom." *American Philosophical Quarterly* 20:277–86.

Knoch, D., A. Pascual-Leone, K. Meyer, V. Treyer, and E. Fehr. 2006. "Diminishing Reciprocal Fairness by Disrupting the Right Prefrontal Cortex." *Science* 314:829–32.

Koenigs, M., L. Young, R. Adolphs, D. Tranel, F. Cushman, M. Hauser, and A. Damasio. 2007. "Damage to the Prefrontal Cortex Increases Utilitarian Moral Judgements." *Nature* 446:908–11.

Laibson, D. I. 1997. "Golden Eggs and Hyperbolic Discounting." *Quantitative Journal of Economics* 112:443–77.

Lakoff, George. 2008. *The Political Mind: Why You Can't Understand 21st-Century American Politics with an 18th-Century Brain.* New York: Viking.

Lamy, M. 2007. "For Juice or Money: The Neural Response to Intertemporal Choice of Primary and Secondary Rewards." *Journal of Neuroscience* 27 (45):12121–22.

Lao-Tzu. 1994. *Tao Te Ching.* Trans. D. C. Lau. New York: Everyman's Library/Knopf.

Lasch, Christopher. 1979. *The Culture of Narcissism: American Life in an Age of Diminishing Expectations.* New York: Norton.

Lattman, P., and J. Strasburg. 2008. "Bear's Fall Sparks Soul Searching." *Wall Street Journal,* March 18, p. C1.

Lawton, C. 2007. "Understanding What You Know: How Business Intelligence Has Come of Age." *Wall Street Journal,* October 23, p. B2.

LeDoux, Joseph. 1996. *The Emotional Brain: The Mysterious Underpinnings of Emotional Life.* New York: Simon & Schuster.

Lehrer, Jonah. 2008. "A New State of Mind." *Seed,* July/August, pp. 64–72.

———. 2009. *How We Decide.* Boston: Houghton Mifflin Harcourt.

Lehrer, Keith, B. Jeannie Lum, Beverly A. Slichta, and Nicholas D. Smith, eds. 1996. *Knowledge, Teaching and Wisdom.* Dordrecht: Kluwer Academic.

Lincoln, Abraham. 1989. *Speeches and Writings, 1859–1865: Speeches, Letters, and Miscellaneous Writings. Presidential Messages and Proclamations.* New York: Library of America.

Lomranz, J., ed. 1998. *Handbook of Aging and Mental Health: An Integrative Approach.* New York: Plenum.

Lutz, A., J. Brefczynski-Lewis, T. Johnstone, and R. J. Davidson. 2008. "Regulation of the Neural Circuitry of Emotion by Compassion Meditation: Effects of Meditative Expertise." *PloS One* 3:e1897.

Lutz, A., L. L. Greischar, N. B. Rawlings, M. Ricard, and R. J. Davidson. 2004. "Long-Term Meditators Self-Induce High-Amplitude Gamma Synchrony During Mental Practice." *Proceedings of the National Academy of Sciences* 101:16,369–73.

Lyons, D. M., H. Afarian, A.-F. Schatzberg, A. Sawyer-Glover, and M. M. Moseley. 2002. "Experience-Dependent Asymmetric Variation in Primate Prefrontal Morphology." *Behavioral Brain Research* 136:51–59.

Mahncke, H. W., B. B. Connor, J. Appelman, O. N. Ahsanuddin, J. L. Hardy, R. A. Wood, N. M. Joyce, T. Boniske, S. M. Atkins, and M. M. Merzenich. 2006. "Memory Enhancement in Older Healthy Adults Using a Brain Plasticity-Based Training Program: A Randomized, Controlled Study." *Proceedings of the National Academy of Sciences* 103: 12,523–28.

Marcus Aurelius, 1992. *Meditations.* Trans. A. S. L. Farquharson. New York: Everyman's Library/Knopf.

Maremont, M. 2007. "Scholars Link Success of Firms to Lives of CEOs." *Wall Street Journal,* September 5, p. A1.

Mather, M., T. Canli, T. English, S. Whitfield, P. Wais, K. Ochsner, J. D. E. Gabrieli, and L. L. Carstensen. 2004. "Amygdala Responses to Emotionally Valenced Stimuli in Older and Younger Adults." *Psychological Science* 15:259–63.

Matson, Wallace I. 1987a. *A New History of Philosophy.* Vol. I: *Ancient & Medieval.* San Diego: Harcourt Brace Jovanovich.

———. 1987b. *A New History of Philosophy.* Vol. II: *Modern.* San Diego: Harcourt Brace Jovanovich.

———. 2006. *Uncorrected Papers: Diverse Philosophical Dissents.* Amherst, NY: Humanity Books.

McClure, S. M., K. M. Ericson, D. I. Laibson, G. Loewenstein, and J. D. Cohen. 2007. "Time Discounting for Primary Rewards." *Journal of Neuroscience* 27 (21):5796–804.

McClure, S. M., D. I. Laibson, G. Loewenstein, and J. Cohen. 2004. "Separate Neural Systems Value Immediate and Delayed Monetary Rewards." *Science* 306:503–7.

McEwen, Bruce S. 2006. "Protective and Damaging Effects of Stress Mediators: Central Role of the Brain." *Dialogues in Clinical Neurosciences* 8:367–81.

McGregor, J. 2007. "The Business Brain in Close-up." *BusinessWeek,* July 23.

McRae, K., K. N. Ochsner, I. B. Mauss, J. J. D. Gabrieli, and J. J. Gross. 2008. "Gender Differences in Emotion Regulation: An fMRI Study of Cognitive Reappraisal." *Group Processes & Intergroup Relations* 11 (2):143–62.

Meacham, J. A. 1990. "The Loss of Wisdom." In Sternberg 1990, pp. 181–211.

Medawar, Peter. 1984. *Pluto's Republic: Incorporating "The Art of the Soluble" and "Induction and Intuition in Scientific Thought."* New York: Oxford University Press.

Meeks, T. W., and D. V. Jeste. 2009. "Neurobiology of Wisdom." *Archives of General Psychiatry* 66 (4):355–65.

Mencius. 1970. *Mencius.* Trans. D. C. Lau. New York: Penguin.

Michaelis, David. 2007. *Schulz and Peanuts: A Biography.* New York: Harper.

Milinski, M., and B. Rockenbach. 2008. "Punisher Pays." *Nature* 452:297–98.

Miller, E. K., and J. D. Cohen. 2001. "An Integrative Theory of Prefrontal Cortex Function." *Annual Review of Neuroscience* 24:167–202.

Miller, G. 2008a. "The Roots of Morality." *Science* 320:734–27.

———. 2008b. "Growing Pains for fMRI." *Science* 320:1412–14.

Moll, J., and R. de Oliveira-Souza. 2007. "Response to Greene: Moral Sentiments and Reason: Friends or Foes?" *Trends in Cognitive Sciences* 11 (8):323–24.

Moll, J., R. de Oliveira-Souza, G. J. Garrido, I. E. Bramati, E. M. A. Caparelli-Daquer, M. L. M. F. Pavia, R. Zahn, and J. Grafman. 2007. "The Self as a Moral Agent: Linking the Neural Bases of Social Agency and Moral Sensitivity." *Social Neuroscience* 2 (3):336–52.

Moll, J., R. Zahn, R. de Oliveira-Souza, F. Krueger, and J. Grafman. 2005. "The Neural Basis of Human Moral Cognition." *Nature Reviews Neuroscience* 6:799–809.

Montague, P. R. 2008. "Free Will." *Current Biology* 18 (4):R584–85.

Montague, P. R., P. Dayan, C. Person, and T. J. Sejnowski. 2002. "Bee Foraging in Uncertain Environments Using Predictive Hebbian Learning." *Nature* 377:725–28.

Montague, P. R., S. E. Hyman, and J. D. Cohen. 2004. "Computational Roles for Dopamine in Behavioural Control." *Nature* 431:760–67.

Montague, Read. 2006. *Why Choose This Book? How We Make Decisions.* New York: Dutton.

Montaigne, Michel de. 2003. *The Complete Works.* Trans. Donald Frame. New York: Everyman's Library/Knopf.

Morante, Elsa. 1977. *History: A Novel.* Trans. William Weaver. New York: Knopf.

Murphy, Cullen. 1998. *The World According to Eve: Women and the Bible in Ancient Times and Our Own.* Boston: Houghton Mifflin.

New Oxford Annotated Bible (Augmented Third Edition). 2007. Ed. Michael D. Coogan. New York: Oxford University Press.

Nielsen, L., B. Knutson, and L. L. Carstensen. 2008. "Affect Dynamics, Affective Forecasting, and Aging." *Emotion* 8 (3):318–30.

Nocera, Joe. 2008. *Good Guys and Bad Guys: Behind the Scenes with the Saints and Scoundrels of American Business (and Everything in Between).* New York: Portfolio/Penguin.

Norenzayan, A., and A. F. Shariff. 2008. "The Origin and Evolution of Religious Prosociality." *Science* 322:58–62.

Nowak, M. A. 2006. "Five Rules for the Evolution of Cooperation." *Science* 314:1560–63.

Nozick, Robert. 1989. *The Examined Life: Philosophical Meditations.* New York: Simon & Schuster.

Nuland, Sherwin B. 1994. *How We Die: Reflections on Life's Final Chapter.* New York: Knopf.

Ochsner, K. N., and J. J. Gross. 2008. "Cognitive Emotion Regulation: Insights from Social Cognitive and Affective Neuroscience." *Current Directions in Psychological Science* 17:153–58.

Parker, K. J., C. L. Buckmaster, K. R. Justus, A. F. Schatzberg, and D. M. Lyons. 2005. "Mild Early Life Stress Enhances Prefrontal-Dependent Response Inhibition in Monkeys." *Biological Psychiatry* 57:848–55.

Parker, K. J., C. Buckmaster, A. F. Schatzberg, and D. M. Lyons. 2004. "Prospective Investigation of Stress Inoculation in Young Monkeys." *Archives of General Psychiatry* 61:933–41.

Parker, K. J., C. L. Buckmaster, K. Sundlass, A. F. Schatzberg, and D. M. Lyons. 2006. "Maternal Mediation, Stress Inoculation, and the Development of Neuroendocrine Stress Resistance in Primates." *Proceedings of the National Academy of Sciences* 103:3000–3005.

Pasupathi, M., U. M. Staudinger, and P. B. Baltes. 2001. "Seeds of Wisdom: Adolescents' Knowledge and Judgment About Difficult Life Problems." *Developmental Psychology* 37:351–61.

Paulhus, D. L., P. Wehr, P. D. Harms, and D. I. Strasser. 2002. "Use of Exemplar Surveys to Reveal Implicit Types of Intelligence." *Personality and Social Psychology Bulletin* 28 (7):1051–62.

Pausch, Randy, with Jeffrey Zaslow. 2008. *The Last Lecture.* New York: Hyperion.

Pfaff, Donald W. 2007. *The Neuroscience of Fair Play: Why We (Usually) Follow the Golden Rule.* New York: Dana Press.

Pinker, Steven. 2002. *The Blank Slate: The Modern Denial of Human Nature.* New York: Viking.

———. 2008. "The Moral Instinct." *New York Times Magazine,* January 13.

Plato. 2000. *Symposium and Phaedrus.* Trans. Tom Griffith. New York: Everyman's Library/ Knopf.

———. 2003. *The Last Days of Socrates: Eythyphro, Apology, Crito, Phaedo.* Trans. Hugh Tredennick and Harold Tarrant. New York: Penguin.

Plutarch. 1960. *The Rise and Fall of Athens: Nine Greek Lives.* Trans. Ian Scott-Kilvert. New York: Penguin.

Pojman, Louis P., ed. 1998. *Classics of Philosophy.* New York: Oxford University Press.

Pomeroy, Sarah B., Stanley M. Burstein, Walter Donlan, and Jennifer Tolbert Roberts. 1999. *Ancient Greece: A Political, Social, and Cultural History.* New York: Oxford University Press.

Posner, Michael I., and Marcus E. Raichle. 1997. *Images of Mind.* New York: Scientific American Library.

Poundstone, William. 1992. *Prisoner's Dilemma.* New York: Doubleday.

Quervain, D. J.-F. de, U. Fischbacher, V. Treyer, M. Schellhammer, U. Schnyder, A. Buck, and E. Fehr. 2004. "The Neural Basis of Altruistic Punishment." *Science* 305:1254–58.

Rachlin, Howard, and David I. Laibson, eds. 1997. *The Matching Law: Papers in Psychology and Economics by Richard J. Herrnstein.* Cambridge, Mass.: Harvard University Press.

Raskin, R. N., and C. S. Hall. 1979. "A Narcissism Personality Inventory." *Psychological Reports* 45:590.

Redish, A. D. 2004. "Addiction as a Computational Process Gone Awry." *Science* 306:1944–47.

Revel, Jean-François, and Matthieu Ricard. 1999. *The Monk and the Philosopher: A Father and Son Discuss the Meaning of Life.* Trans. John Canti. New York: Schocken.

Ricard, Matthieu. 2006. *Happiness: A Guide to Developing Life's Most Important Skill.* Boston: Little, Brown.

Rice, Eugene F., Jr. 1958. *The Renaissance Idea of Wisdom.* Westport, Conn.: Greenwood Press.

Richerson, Peter J., and Robert Boyd. 2005. *Not By Genes Alone: How Culture Transformed Human Evolution.* Chicago: University of Chicago Press.

Ridley, Matt. 1996. *The Origins of Virtue: Human Instincts and The Evolution of Cooperation.* New York: Viking.

Rilling, J. K., D. A. Gutman, T. R. Zeh, G. Pagnoni, G. S. Berns, and C. D. Kilts. 2002. "A Neural Basis for Social Cooperation." *Neuron* 35:395–405.

Robinson, D. N. 1990. "Wisdom Through the Ages." In Sternberg 1990, pp. 13–24.

Rochat, M. J., E. Serra, L. Fadiga, and V. Gallese. 2008. "The Evolution of Social Cognition: Goal Familiarity Shapes Monkeys' Action Understanding." *Current Biology* 18:1–6.

Rooney, Dan (as told to Andrew E. Masich and David F. Halaas). 2007. *Dan Rooney: My 75 Years with the Pittsburgh Steelers and the NFL.* New York: Da Capo.

Rosati, A. G., J. R. Stevens, B. Hare, and M. D. Hauser. 2007. "The Evolutionary Origins of Human Patience: Temporal Preferences in Chimpanzees, Bonobos, and Human Adults." *Current Biology* 17:1663–68.

Rosen, C. 2008. "The Myth of Multitasking." *The New Atlantis* 20: 105–10.

Rosenbloom, S. 2008. "Generation Me vs. You Revisited." *New York Times,* January 17, p. G1.

Rowatt, W. C., C. Powers, V. Targhetta, J. Comer, S. Kennedy, and J. Labouf. 2006. "Development and Initial Validation of an Implicit Measure of Humility Relative to Arrogance." *Journal of Positive Psychology* 1 (4):198–211.

Rozin, P., J. Haidt, and K. Fincher. 2009. "From Oral to Moral." *Science* 323:1179–80.

Samanez-Larkin, G. R., N. G. Hollon, L. L. Carstensen, and B. Knutson. 2008. "Individual Differences in Insular Sensitivity During Loss Anticipation Predict Avoidance Learning." *Psychological Science* 19 (4):320–23.

Sample, Susan. 2007. "A Nobel Effort." *Continuum: The Magazine of the University of Utah* 17 (3); http://www.continuum.utah.edu/2007winter/feature_3.html. (This is an updated version of a story that originally ran in the Winter 1997 issue of the *University of Utah Health Sciences Report.*)

Sanfey, A. G., and L. J. Chang. 2008. "Multiple Systems in Decision Making." *Annals of the New York Academy of Sciences* 1128:53–62.

Sanfey, A. G., J. K. Rilling, J. A. Aronson, L. E. Nystrom, and J. D. Cohen. 2003. "The Neural Basis of Economic Decision-Making in the Ultimatum Game." *Science* 300:1755–58.

Schaie, K. Warner, and James E. Birren, eds. 2001. *Handbook of the Psychology of Aging.* 5th ed. San Diego: Academic Press.

Schiller, D., I. Levy, Y. Niv, J. E. LeDoux, and E. A. Phelps. 2008. "From Fear to Safety and Back: Reversal of Fear in the Human Brain." *Journal of Neuroscience* 28:11,517–25.

Schnall, S., J. Haidt, G. L. Clore, and A. H. Jordan. 2008. "Disgust as Embodied Moral Judgment." *Personality and Social Psychology Bulletin* 34:1096–1109.

Seneca. 1962. *Ad Lucilium epistulae morales.* Trans. Richard M. Gummere. Cambridge, Mass.: Harvard University Press.

Seymour, B., T. Singer, and R. Dolan. 2007. "The Neurobiology of Punishment." *Nature Reviews Neuroscience* 8:300–311.

Shafir, S., T. Reich, E. Tsur, I. Erev, and A. Lotem. 2008. "Perceptual Accuracy and Conflicting Effects of Certainty on Risk-Taking Behaviour." *Nature* 453:917–20.

Sharot, T., A. M. Riccardi, C. M. Raio, and E. A. Phelps. 2007. "Neural Mechanisms Mediating Optimism Bias." *Nature* 450:102–5.

Shenk, Joshua Wolf. 2005. *Lincoln's Melancholy: How Depression Changed a President and Fueled His Greatness.* Boston: Houghton Mifflin.

———. 2009. "What Makes Us Happy?" *The Atlantic,* June, pp. 36–53.

Singer, T. 2006. "The Neuronal Basis and Ontogeny of Empathy and Mind Reading: Review of Literature and Implications for Future Research." *Neuroscience and Biobehavioral Reviews* 30:855–63.

Singer, T., B. Seymour, J. O'Doherty, H. Kaube, R. J. Dolan, and C. D. Frith. 2004. "Empathy for Pain Involves the Affective but Not Sensory Components of Pain." *Science* 303:1157–62.

Singer, T., B. Seymour, J. P. O'Doherty, K. E. Stephan, R. J. Dolan, and C. D. Frith. 2006. "Empathic Neural Responses Are Modulated by the Perceived Fairness of Others." *Nature* 439:466–69.

Slagter, H. A., A. Lutz, L. L. Greischar, A. D. Francis, S. Nieuwenhuis, J. M. Davis, and R. J. Davidson. 2007. "Mental Training Affects Distribution of Limited Brain Resources." *PloS Biology* 5:1228–35.

Smith, Adam. 1976. *The Theory of Moral Sentiments.* Ed. D. D. Raphael and A. L. Macfie. Oxford: Clarendon Press.

Smith, Jacqui. 2007. "The Allure of Wisdom." *Human Development* 50:367–70.

Sokol-Hessner, P., M. Hsu, N. G. Curley, M. R. Delgado, C. F. Camerer, and E. A. Phelps. 2009. "Thinking Like a Trader Selectively Reduces Individuals' Loss Aversion." *Proceedings of the National Academy of Sciences* 106:5035–40.

Soon, C. S., M. Brass, H.-J. Heinze, and J.-D. Haynes. 2008. "Unconscious Determinants of Free Decisions in the Human Brain." *Nature Neuroscience* 11:543–45.

Spiers, H. J., and E. A. Maguire. 2008. "The Dynamic Nature of Cognition During Wayfinding." *Journal of Environmental Psychology* 28:232–49.

Staudinger, U. M. 1999. "Older and Wiser? Integrating Results on the Relationship Between Age and Wisdom-Related Performance." *International Journal of Behavioral Development* 23:641–64.

Staudinger, U. M., and P. B. Baltes. 1996. "Interactive Minds: A Facilitative Setting for Wisdom-Related Performance?" *Journal of Personality and Social Psychology* 71:746–62.

Staudinger, U. M., J. Domer, and C. Mickler. 2005. "Wisdom and Personality." In Sternberg and Jordan 2005, pp. 191–219.

Sternberg, R. J. 1985. "Implicit Theories of Intelligence, Creativity, and Wisdom." *Journal of Personality and Social Psychology* 49:607–27.

Sternberg, Robert J., ed. 1990. *Wisdom: Its Nature, Origins, and Development.* New York: Cambridge University Press.

Sternberg, Robert J., and Todd I. Lubart. 2001. "Wisdom and Creativity." In Schaie and Birren 2001, pp. 500–522.

Sternberg, Robert J., and Jennifer Jordan, eds. 2005. *A Handbook of Wisdom: Psychological Perspectives.* New York: Cambridge University Press.

Sternberg, R. J., A. Reznitskaya, and L. Jarvin. 2007. "Teaching for Wisdom: What Matters Is Not Just What Students Know, But How They Use It." *London Review of Education* 5:143–58.

Stevens, J. R., E. V. Hallinan, and M. D. Hauser. 2005. "The Ecology and Evolution of Patience in Two New World Monkeys." *Biology Letters* 1 (2):223–26.

Stix, G. 2008. "Of Survival and Science." *Scientific American,* October 6.

Stone, I. F. 1988. *The Trial of Socrates.* New York: Little, Brown.

Strathern, Paul. 1999. *Confucius in 90 Minutes.* Chicago: Ivan R. Dee.

Strong, John S. 2001. *The Buddha: A Short Biography.* Oxford: Oneworld.

———. 2002. *The Experience of Buddhism: Sources and Interpretations.* 2nd ed. Belmont, Calif.: Wadsworth/Thomson Learning.

Sulzberger, A. G., and S. Chan. 2009. "Despite Outcry, Gandhi's Meager Belongings Sell for $1.8 Million." *New York Times,* March 6, p. A1.

Surowiecki, James. 2004. *The Wisdom of Crowds.* New York: Doubleday.

Talmi, D., and C. Frith. 2007. "Feeling Right About Doing Right." *Nature* 446:865–66.

Tang, A., B. C. Reeb, R. D. Romeo, and B. S. McEwen. 2003. "Modification of Social Memory, Hypothalamic-Pituitary-Adrenal Axis, and Brain Asymmetry by Neonatal Novelty Exposure." *Journal of Neuroscience* 23:8254–60.

Tangney, J. P. 2000. "Humility: Theoretical Perspectives, Empirical Findings and Directions for Future Research." *Journal of Social and Clinical Psychology* 19:70–82.

———. 2002. "Humility." In *Handbook of Positive Psychology,* ed. C. R. Snyder and S. J. Lopez, pp. 411–19. Oxford: Oxford University Press.

Telushkin, Joseph. 1994. *Jewish Wisdom: Ethical, Spiritual, and Historical Lessons from the Great Works and Thinkers.* New York: William Morrow.

Templeton, John Marks. 1997. *Worldwide Laws of Life.* Philadelphia: Templeton Foundation Press.

Thaler, Richard H., and Cass R. Sunstein. 2008. *Nudge: Improving Decisions About Health, Wealth, and Happiness.* New Haven, Conn.: Yale University Press.

Thorn, John, and Pete Palmer, eds. 1989. *Total Baseball.* New York: Warner.

Tuchman, B. W. 1979. "An Inquiry into the Persistence of Unwisdom in Government." *Parameters: Journal of the U.S. Army War College* 10 (1):2–9.

Urry, H. L., J. B. Nitschke, I. Dolski, D. C. Jackson, K. M. Dalton, C. J. Mueller, M. A. Rosenkranz, C. D. Ryff, B. H. Singer, and R. J. Davidson. 2004. "Making a Life Worth Living: Neural Correlates of Well-being." *Psychological Science* 15:367–72.

Urry, H. L., C. M. van Reekum, T. Johnstone, N. H. Kalin, M. E. Thurow, H. S. Schaefer, C. A. Jackson, C. J. Frye, L. L. Greischar, A. L. Alexander, and R. J. Davidson. 2006. "Amygdala and Ventromedial Prefrontal Cortex Are Inversely Coupled During Regulation of Negative Affect and Predict the Diurnal Pattern of Cortisol Secretion Among Older Adults." *Journal of Neuroscience* 26:4415–25.

Vaillant, George E. 1977. *Adaptation to Life.* Boston: Little, Brown.

———. 1993. *The Wisdom of the Ego.* Cambridge, Mass.: Harvard University Press.

———. 2003. *Aging Well: Surprising Guideposts to a Happier Life from the Landmark Harvard Study of Adult Development.* New York: Little, Brown.

Vaillant, George E., J. Templeton, M. Ardelt, and S. E. Meyer. 2008. "The Natural History of Male Mental Health: Health and Religious Involvement." *Social Science & Medicine* 66:221–31.

Van Doren, Carl. 1938. *Benjamin Franklin.* New York: Viking.

Vignemont, F. de, and T. Singer. 2006. "The Empathic Brain: How, When and Why?" *Trends in Cognitive Sciences* 10:435–41.

Wallis, Glenn. 2007. *Basic Teachings of the Buddha: A New Translation and Compilation, with a Guide to Reading the Texts.* New York: Modern Library/Vintage.

Watson, T. L., and F. Blanchard-Fields. 1998. "Thinking with Your Head *and* Your Heart: Age Differences in Everyday Problem-Solving Strategy Preferences." *Aging, Neuropsychology, and Cognition* 5:225–40.

Weaver, I. C. G., N. Cervoni, F. A. Champagne, A. C. D'Alessio, S. Sharma, J. R. Seckl, S. Dymov, M. Szyf, and M. J. Meaney. 2004. "Epigenetic Programming by Maternal Behavior." *Nature Neuroscience* 7:847–54.

Westen, Drew. 2007. *The Political Brain: The Role of Emotion in Deciding the Fate of the Nation.* New York: PublicAffairs.

Wheatley, T., and J. Haidt. 2005. "Hypnotic Disgust Makes Moral Judgments More Severe." *Psychological Science* 16:780–84.

Wiesel, Elie. 2003. *Wise Men and Their Tales: Portraits of Biblical, Talmudic, and Hasidic Masters.* New York: Schocken.

Wilde, Oscar. 1931. *The Writings of Oscar Wilde.* Vol. 5: *Social Reform.* New York: Wm. H. Wise.

Wills, Garry. 1992. *Lincoln at Gettysburg: The Words That Remade America.* New York: Simon & Schuster.

Wilson, Edward O. 1994. *Naturalist.* Washington, D.C.: Island Press/Shearwater Books.

Wright, Robert. 1994. *The Moral Animal: Why We Are the Way We Are: The New Science of Evolutionary Psychology.* New York: Pantheon.

Yang, C., P. Belawat, E. Hafen, L. Y. Jan, and Y. Jan. 2008. "Drosophila Egg-Laying Site Selection as a System to Study Simple Decision-Making Processes." *Science* 319:1679–83.

Young, L., F. Cushman, M. Hauser, and R. Saxe. 2007. "The Neural Basis of the Interaction Between Theory of Mind and Moral Judgment." *Proceedings of the National Academy of Sciences* 104:8235–40.

INDEX

Abishag, 149
Abraham, 44
acetylcholine, 204
ACTH, 219
action bias, 8
Adaptation to Life (Vaillant), 41, 172, 233–7
addictions, 171, 173, 184, 199
Adonijah, 148–9, 156–7
adversity, early-life, 211–26
  examples of, 211–15, 217, 225–6
  seeds of wisdom sown by, 216–17
  stress-inoculation research on, 218–22,
    223, 224
  see also emotional resilience
Aesop, 38, 182–3, 184, 187–9, 294n–5n
affective forecasting, 71
aging, 62, 67–78, 108, 202, 227–40
  affective forecasting in, 71
  brain changes of, 230–2
  cognitive declines of, 63, 77, 228, 229–32,
    238
  cognitive exercise for, 231–2
  coping mechanisms in, 234–7
  emotional intelligence of, 228–9
  emotional regulation and, 220, 223, 229,
    238, 239–40, 255
  grandparent hypothesis of, 78
  "let it be" approach of, 238–9
  loss associated with, 69
  problem-solving skills in, 223, 227–9,
    232, 238, 239
  psychological changes in, 232–7
  social cognition in, 73, 223, 228, 238–9
  social isolation and, 67–8

temporal discounting and, 72, 90, 184
time horizons of, 41, 51–2, 63–4, 68–9,
    70–2, 223
aging, research on, 39–48, 53–6, 67–73,
    228–37, 239–40
  by gerontologists, 41–2, 67
  lawyers in, 45–7
  longitudinal studies in, 232–7
  in nursing homes, 42, 67–8
  wisdom literature in, 42–7, 50
  see also life-span development
Aging Well (Vaillant), 237
Ainslie, George, 169, 171, 172–7, 178, 179,
    180, 181, 185, 187–9, 200, 201, 233–4,
    235, 260
Allen, Sarah, 166
Alter, Robert, 44
altruism, 16, 19, 34, 35, 95, 114, 129, 143,
    147–68, 199, 235–6
  of bees, 151, 271
  civic, 30
  community reputation enhanced by, 154,
    155
  evolution of, 150–2, 155, 157, 245
  Golden Rule and, 152
  neurobiology of, 147, 148, 153–6, 162–3,
    166–7
  punitive aspect of, *see* punishment
  reciprocal, 151–2
  social benefits of, 159–61
  see also cooperation; social justice
Amodio, David M., 261
Anacharsis, 257
Analects (Confucius), 30, 61, 196, 250–1

PERMISSIONS ACKNOWLEDGMENTS

Grateful acknowledgment is made to the following for permission to reprint previously published material:

American Economic Association: Excerpts from "The Vulcanization of the Human Brain: A Neural Perspective on Interactions Between Cognition and Emotion" by Jonathan D. Cohen (*Journal of Economic Perspectives,* volume 19, number 4, Fall 2005). Reprinted by permission of the American Economic Association.

Penguin Group (UK): Excerpts from *The Last Days of Socrates* by Plato, translated by Hugh Tredennick, copyright © 1954, 1959, 1969 by Hugh Tredennick (Penguin Classics, second revised edition, 1969). Reprinted by permission of Penguin Group (UK).

A NOTE ABOUT THE AUTHOR

Stephen S. Hall is the author of five previous books about contemporary science, including *Invisible Frontiers* (a narrative account of the birth of the biotechnology industry) and *Merchants of Immortality* (on stem cell science and the birth of regenerative medicine). He is a regular contributor to *The New York Times Magazine, National Geographic,* and many other publications. This book grew out of a 2007 cover story for *The New York Times Magazine* titled "Can Science Tell Us Who Grows Wiser?" In addition to writing, Hall teaches in the journalism programs at Columbia University and New York University. He lives in Brooklyn, New York, with his wife and two children.

A NOTE ON THE TYPE

This book was set in Adobe Garamond. Designed for the Adobe Corporation
by Robert Slimbach, the fonts are based on types first cut by Claude Garamond (c. 1480–1561).

Composed by Creative Graphics, Allentown, Pennsylvania
Printed and bound by Berryville Graphics, Berryville, Virginia
Book design by Robert C. Olsson